HISTORY OF LUNAR EXPLORATION

HISTORY OF LUNAR EXPLORATION

Robert Reeves

FIREFLY BOOKS

A Firefly Book

Published by Firefly Books Ltd. 2026

First printing

Library of Congress Control Number: 2026938644

Library and Archives Canada Cataloguing in Publication
Title: History of lunar exploration / Robert Reeves.
Names: Reeves, Robert, 1946- author.
Description: Includes index.
Identifiers: Canadiana 20260181013 | ISBN 9780228106289 (hardcover)
Subjects: LCSH: Space flight to the Moon—History. |
LCSH: Moon—Exploration—History.
Classification: LCC TL799.M6 R44 2026 | DDC 629.45/4—dc23

Published in the United States by
Firefly Books (U.S.) Inc.
P.O. Box 1338, Ellicott Station
Buffalo, New York
14205

Published in Canada by
Firefly Books Ltd.
50 Staples Avenue, Unit 1
Richmond Hill, Ontario
L4B 0A7

Cover and interior design: Stacey Cho

Printed in China | E

To Anne, whose love makes me embrace the future.

Table of Contents

Preface

For me, one of the most amazing aspects of thinking and writing about the space program's push toward the Moon was the realization that I was alive during the early heated, politically charged Space Race. This stunning period in history, when the battle for technological supremacy between the United States and the USSR was played out with space rockets and satellites and the drama of each new mission was broadcast to everyone's living room, did not take place very long ago.

I was in the fifth grade when Sputnik turned the world upside down and announced the Space Age had arrived. America's Pioneer and the USSR's early Luna Moon probes dominated the headlines when I was in grade school. America's Ranger spacecraft successfully reached the Moon while I was in high school. I was an adult by the time the National Aeronautics and Space Administration's (NASA) Surveyor spacecraft landed on the Moon, the Lunar Orbiter probes charted the Moon in detail, and the dramatic landing of Apollo 11 ended the political race to the Moon. I and millions more who are still alive today witnessed these remarkable achievements first-hand.

Sadly, just as Apollo's 1969 lunar landing triumph established America as the technological leader of the world, a shortsighted Congress ended the Moon program and further lunar exploration was non-existent until the near end of the century. So abrupt was the termination of the Apollo program that already built and paid for Saturn V rockets and spacecraft for Apollo 18, 19, and 20 became incredibly expensive museum exhibits.

A 16th-century parallel to the heartbreakingly senseless termination of the Apollo program would be if Europe abandoned exploration after Columbus discovered the New World and, by royal decree, all shipbuilding blueprints would be destroyed and all shipyards burned to the ground to prevent any further explorations.

Fortunately, in the 21st century, momentum is building for a return to the Moon, this time with international partners. Despite the more than 50-year hiatus in manned lunar landings, I may live long enough to again see fresh footprints on the Moon. NASA's Artemis program shows promise of returning humans to the Moon within the 2020s. Like someone who never let the flame of an old high school sweetheart fade away, I will be watching and waiting to embrace Luna once again and savor the triumph of humankind's return to the Moon.

Introduction

At the dawn of the Space Age in the 1960s, very little was known about the Moon despite three centuries of telescopic study since Galileo first pointed his lens at the Moon. To scout the Moon prior to the manned Apollo landings, three series of American spacecraft, known as Ranger, Surveyor, and Lunar Orbiter, gathered data that greatly increased our knowledge about the Moon. The Soviets also sent their Luna series of probes to the Moon, motivated by the chance of a political and scientific victory.

In September 1959, Luna 2 became the first human-made object to reach the lunar surface. A month later, Luna 3 finally revealed the far side of the Moon to humanity. These Soviet lunar triumphs ignited an American drive to win the Space Race, culminating with Neil Armstrong's amazing "giant leap for mankind" on the plains of Mare Tranquillitatis a mere 10 years after Luna 2 reached the Moon. Although these missions were flown more than half a century ago, their data continues to provide modern researchers with insights into the physical structure of the Moon.

In the 1960s, the most important piece of information supplied by the robotic predecessors to Apollo was that the Moon's surface, though dusty, was firm enough to support a spacecraft and a human walking on it. In the politically charged Cold War race to reach the Moon, this single piece of information was key to allowing Apollo to take the risk of landing humans on the Moon. The five successful Surveyor landings along with two Soviet landers forever banished Tommy Gold's dire warning that portions of the Moon were covered with heaps of quicksand-like dust that would spell doom for any landing craft. Before Apollo set a single boot print on the Moon, robots had mapped the Moon at resolutions 10 times or more than what was possible from Earth. The Moon machines also confirmed the volcanic origin of the dark mare areas that trace the Man-in-the-Moon on the lunar face.

The Surveyor probes also proved that the chemical composition of the Moon is far from the pristine primordial undifferentiated ball of rock it was thought to be. But in the bigger picture of understanding the Moon, it was the Lunar Orbiters that provided grand insights into the creation of the lunar surface we see today. Lunar Orbiter 1, 2, and 3 succeeded in finding safe but scientifically interesting landing sites for Apollo. Lunar Orbiter 4 took the spectacular image of the Orientale basin that provided the key to complete our understanding of the impact origins of most lunar features. Lunar Orbiter 5 made detailed mosaics of the fresh craters Aristarchus, Copernicus, and Tycho, as well as volcanic areas like the Marius Hills, and provided insights about the geology of the Moon.

For researchers, going from near ignorance of the Moon's physical properties in 1959 to fully mapping its entire globe and having astronauts walk on the surface within 10 years was akin to scientific whiplash. The torrent of information was difficult to fully analyze in the hurried frenzy to land humans on the Moon before the 1960s were over. Then, almost as fast as the lunar gold rush had started, it was over. Once the first politically driven manned Apollo landings were completed, a myopic Congress was all too happy to scrap future lunar exploration and apply space money to war and pork-barrel projects that achieved nothing of consequence.

When the headlong rush to land men on the Moon cooled off, there was time to digest what had been accomplished. For the next quarter-century, no scientific spacecraft from Earth approached our Moon. The data gathered at breakneck speed in the 1960s could now be analyzed and reduced into an understanding of the world that we so desperately sought to reach, then abandoned once we got there.

1

Earthbound Explorers

For millennia humankind has watched and pondered the silvery orb we call the Moon. The celestial wanderings of Earth's natural satellite fascinated the curious, terrified the superstitious and inspired reverence as a god by many ancient cultures. The lure of the Moon entwined its pale glow with cultural mythology, volumes of literature and prose, and an insatiable desire to understand what the Moon was and why it perpetually crossed the night sky.

The slow but methodical march to understanding the Moon traverses the lives of some of humankind's greatest thinkers. Those who pondered the Moon from ancient times up to the birth of the Space Age are rightly regarded as intellectual giants.

Over the course of humanity's time on Earth, the past 400 years are but a blink of an eye on the cosmic time scale. Yet the full sum of our knowledge about the Moon comes from the period since the telescope was perfected. More amazing is how scientifically ignorant we remained about the Moon for the first 350 years of those four centuries.

The maturation of rocket technology in the mid-20th century coincided with the political Cold War struggle between Western democracies and the Communist world led by the USSR. That political clash spawned a technological competition between the two world superpowers, the United States and the Soviet Union, that propelled the 1960s race to the Moon. The triumph of Apollo 11's manned landing on the Moon a mere 12 years after the first Earth satellite, Sputnik, crossed the sky is testament to American engineering genius and industrial might.

Science often rode on the coattails of the politically driven race to the Moon, but one of the lasting legacies of that technological superpower competition was a real understanding of our Moon and its evolution.

The science fiction writer Isaac Asimov once stated that it is unfortunate that Earth has a Moon because ancient sky gazers observed it revolving around the Earth and assumed that the rest of the universe did likewise. This created the erroneous idea of an Earth-centered universe that dominated astronomical thinking until just 400 years ago.

Ancient humans were fascinated by the Moon almost as soon as they could reason and think. There is evidence that as far back as 8,000 years ago European tribes had an interest in the phases of the Moon. In the 1960s, science writer Alexander Marshack became fascinated by patterns of indentations carved into bones dating as far back as 6000 BC. Marshack's controversial conclusion was that the 167 marks engraved on an 8,000-year-old bone were a record of the phases of the Moon over a six-month period. Each of the spots on the weaving line carved into the bone supposedly represented the Moon on a different night. Eventually he found similar markings on bones that were 34,000 years old, indicating there might have been an interest in the Moon since the Stone Age.

The pre-Renaissance view of the Moon was influenced by European and Near Eastern civilizations who saw the Sun, Moon and planets as visible manifestations of their deities. In these cultures, the Moon was almost universally associated with a female goddess. In the Moon, the Egyptians saw their goddess, Isis. The later Greeks viewed the Moon as Artemis, daughter of Zeus and twin sister of Apollo. The earlier Greeks considered the Moon to be Selene, who became the Roman goddesses Diana and Luna.

For a thousand years, the astronomers of Babylonia noticed the Sun, Moon, and wandering planets never departed from a narrow sky-circling band of the twelve constellations we now call the Zodiac. Their astronomical observations concentrated on using the Moon to tell time. After the Assyrians came to power around 745 BC, the emphasis of astronomy shifted from telling time to making astrological predictions and predicting future solar eclipses. They discovered that over a period of 18 years and 11 days, the Moon returned to its starting position in the sky, and thus if the date of one eclipse was known, the occurrence of another could be calculated.

The Greeks began the understanding of how the heavens worked. Indeed, the very word astronomy is derived from ancient Greek: a combination of *aster* or *astron* meaning "star" and *nomos* meaning "law." The Greek scholar Thales, born in 640 BC, taught that the stars were made of fire and the Moon was illuminated by the Sun. Other Greeks, such as Anaximander, were among the first to state that the stars move around the Earth in great circles around the pole star. Anaxagoras taught that the Moon possessed mountains, rivers, lakes, and seas just as Earth does and the habitability of the Moon by humans was an article of philosophical faith.

Pythagoras of Samos, who lived around 500 BC, considered the number 10 to be a perfect number and assumed there were 10 bodies in the universe: the five known planets, plus the Sun, Earth, and Moon made eight, the fixed stars together made nine while a hypothetical "Counter-Earth" brought the total to ten. To locate each of these objects in the sky, Pythagoras embedded them on transparent celestial crystal spheres nested within each other, with the Earth at the center. However flawed, the Greek concept of crystal spheres supporting the heavenly bodies became entrenched in astronomical thinking for the next several thousand years.

Meton of Athens, born in 460 BC, rediscovered the cycle in which the Moon repeats its positions about every 19 years. He found that after 235 lunar months, or 6,940 days, the lunar cycles repeat in what is now known as the Metonic Cycle. This was an important step in understanding the occurrence of eclipses.

Heraclides, one of Plato's disciples, believed the heavenly bodies were solid worlds, each surrounded by an atmosphere, and that the visible spots on the Moon were caused by the fogs that surrounded it. Heraclides' most important contribution was introducing the concept that some heavenly bodies revolve around one another, and that the Earth rotated once every 24 hours, making the stars seem to be circling the Earth.

Aristotle deduced Earth's size in the third century BC by observing Earth's shadow upon the Moon during a lunar eclipse. He reasoned that the Earth's diameter must be approximately three times that of the Moon and the Moon's distance from Earth was 56 Earth radii, figures close to the truth. His calculations used only naked eye observations, but these first hard facts about the Moon began to separate it from myth and mysticism.

Aristarchus of Samos in the second century BC was the first to apply mathematical principles to astronomy. By observing eclipses of the Moon, he proved that the Sun was much larger than the Earth and, consequently, he placed the Sun, not the Earth, at the center of the universe. Seleucus, a student of Aristarchus, was one of the first to recognize tides were connected to the location of the Moon relative to a locale on Earth.

In his studies between 147 and 127 BC, Hipparchus observed the angular diameter of Earth's shadow during a lunar eclipse and compared it to the angular diameter of the Sun and Moon. He reasoned the distance to the Moon must be 59 Earth radii, or 379,800 kilometers, which equates favorably with the modern mean distance to the Moon. He also was the first to note that the Moon's relative motion around Earth was not circular and that the inclination of the Moon's orbit was five degrees to Earth's equator.

After Hipparchus, progress in astronomy was nearly nonexistent for the next 1,500 years. The only truly noteworthy event occurred in Alexandria when Claudius Ptolemaeus created a multi-volume astronomical text, known today as the *Almagest*, between 127 and 141 AD. More commonly known as Ptolemy, he consolidated all the known astronomical data at that time. However, Ptolemy made a fundamental mistake that doomed astronomical progress for more than a millennium. Ptolemy rejected the Sun-centered universe of Aristarchus and modified the Earth-centered celestial scheme where the Sun, Moon, planets, and stars were embedded on concentric transparent spheres. The Ptolomaic system was flawed, but it could predict the positions of celestial objects to within a degree.

Regarding the Moon, Ptolemy discovered what he called lunar "evection," an observed phenomenon that Hipparchus had previously described as "lunar inequality," that the Moon's orbit was not perfectly circular. Ptolemy discovered that the eccentricity of the Moon's orbit was subject to annual variation dependent upon the motion of the line of apsides, a line drawn through the farthest points of an orbit.

When Alexandria was burned, the combined knowledge of Eastern Mediterranean civilizations housed in the great library went up in smoke. By a curious fortune, some of the few scrolls saved from the destruction were the astronomical works of Ptolemy. His scrolls became the sole source of our knowledge of Greek astronomy and the *Almagest* became the fundamental astronomy text throughout the Middle Ages.

Science Sweeps Away Superstition

In the 1500s, the ages of ignorance came to an abrupt halt. In the span of a little more than a century and a half, a succession of astronomical heroes discovered and defined the true nature of our solar system, banished the Earth-centered view of the universe and laid the foundations for the science of modern astronomy.

The stories of the astronomical heroes of the Renaissance deserve to be repeated in this account of our understanding of the Moon for several reasons. Not only are these people of crucial importance for their groundbreaking scientific achievements, but they also deserve recognition for their perseverance in the face of formidable social, religious and personal barriers. Science will forever thank Nicolaus Copernicus, Tycho Brahe, Johannes Kepler, Galileo Galilei, and Isaac Newton for establishing astronomy as the mother of all sciences. These individuals did little to further our understanding of the Moon itself, but collectively, along with other great thinkers in the 17th century, they played a vital role in increasing our understanding of how the solar system worked and, by default, how the Moon moved in the sky.

Copernicus pioneered the basic order of the solar system, Tycho provided the critical observations Kepler needed to understand the motions of objects in the solar system, Galileo showed the usefulness of the telescope in astronomy and found proof of the Copernican theories. Newton completed the work of Kepler and showed how gravity and inertia keep the Moon and planets moving through our solar system. Without these basic astronomical understandings, not even our close Moon could be fathomed.

While pursuing a doctorate in canon law, Nicolaus Copernicus was intrigued by the astronomy lectures of the mathematician Albert Brudzewski. Copernicus learned Greek to study the original writings of Ptolemy. But the more Copernicus studied the ancient astronomical works, the more he became dissatisfied with the accepted Earth-centered idea of celestial order. He eventually realized the problems of celestial motion associated with the classic Greek theories could be eliminated if the Sun, not the Earth, were placed at the center of the moving planets.

After completing his education, Copernicus took up the duties of canon of the cathedral at Frauenberg. Along with the church duties, he also turned his attention to the skies and introduced practical reforms to the calendar. Eventually, Copernicus produced the most accurate calendar to date, miscalculating the actual length of the year by a mere 28 seconds.

However, Copernicus fell short in his initial efforts to understand the motions of the planets. What Copernicus lacked was an understanding that the planets orbit the Sun in ellipses, not perfect circles. But Copernicus, as a good Christian, could not imagine that God would allow the planets to circle the Sun in anything but a perfect circle.

The early 16th century works of the Polish church Canon Nicolaus Copernicus permanently banished the Dark Ages notion that universe revolved around the Earth.

For more than 30 years, Copernicus worked theoretically and experimentally to prove that the Sun was at the center of the solar system. By 1530 he had fully developed and proven his own heliocentric idea of astronomical order. However, fear of religious persecution prevented him from publishing his then radical ideas. Finally, at age 67, Copernicus agreed to publish his great work. When a copy of the landmark book was finally placed into Copernicus' hands in 1543, he already lay on his deathbed. But the truth revealed in the Copernican system could not be contained by superstition and 1,500 years of stagnation in the Western understanding of the universe came to an end. The order of the Sun, Earth and the planets were at last established and thousands of years of misconception had been swept aside.

Leonardo da Vinci was a Renaissance giant who brought together talents for both art and science. Contemporary assumption was that the dark areas of the Moon were oceans, but he thought that the bright areas of the Moon were seas while the dark areas were land. Leonardo believed that the Moon had an atmosphere and oceans, and because it was covered with so much water, the Moon was a fine reflector of light. He was the first to recognize that the light which sometimes illuminates the dark portion of a crescent Moon is light reflected from the Earth, a phenomenon he called "moon luster," but today is called earthshine.

Mathematics Master the Solar System

The proof that the Copernican system was the proper order in the solar system came from the work of Denmark's Tycho Brahe. Yet, oddly, Tycho rejected the Sun-centered concept of planetary motion. Tycho was greatly impressed by a total solar eclipse of the Sun on October 21, 1560. Two years later at the University of Leipzig he began making astronomical observations and quickly noted that predicted locations of the planets Jupiter and Saturn were in error. Tycho then realized the need for long term systematic observations of the Moon and planets before further theories could be devised regarding their motions.

The late 16th century observations by the Danish astronomer Tycho Brahe provided the data needed by Johanes Kepler to perfect his laws of planetary motion.

In 1572, Tycho saw a star in Cassiopeia flare into unheard of brilliance at a position where no star was previously seen. Over the next 18 months he wrote an exhaustive account of its existence, an object known to this day as "Tycho's Star." His observational accounts of the supernova in *De Stella Nova* brought him fame and he decided to devote the rest of his life to astronomy.

Beginning in 1580, he spent the next 21 years pursuing astronomy at Uraniborg, his private observatory. The primary instrument was the mural quadrant, a huge meridian instrument graduated at 10 arc-second intervals. Using this great measuring device, Tycho's observational work was the most meticulous and precise at that time and he charted the position of 777 stars and the motions of the five visible planets.

Tycho also applied his talents to recording the position of the Moon, and in due course made several discoveries about its motion. He found the Moon had an irregular motion that matched predictions at opposition and conjunction with the Sun, as well as at quadrature, but in the positions between, the Moon might be as far as 40 arc-minutes from its predicted position. This irregularity had been noted in the 10th century, but references to it had been lost. Tycho also found the inclination of the Moon's orbit to the ecliptic was not fixed, but oscillated, and that the motion of the nodes of the Moon's orbit were variable.

Like Copernicus, Johannes Kepler was attracted to both theology and astronomy. However, the lectures of Professor Michael Maestlin convinced Kepler to accept the Copernican theory of cosmology. In 1594 he accepted the Chair of Mathematics at the University of Graz and became engrossed in astronomy that would chart the course of his life.

Although Copernican theory explained the order of the solar system, it did little to explain what drove the planets along their paths around the Sun. Kepler's investigations into this brought him into contact with Galileo and Tycho Brahe. To obtain better observational data to support his calculations, Kepler traveled to Prague to meet Tycho. However, Tycho's observational data did little to help Kepler understand the motion of the planets, especially the orbit of Mars. It wasn't until years later that Kepler realized that Mars moved about the Sun in an ellipse, not a circle. In his computations, Kepler also hit on the idea that it was the Sun which controlled the movement of the planets. This simple concept was won at the expense of the mathematical fight of his life and firmly established the validity of Copernicus' ideas of a Sun-centered universe.

In the early 17th century, the German astronomer Johannes Kepler perfected the three laws of planetary motion that allowed accurate prediction of planetary location.

Kepler's work explained the first two of what became his famous three laws governing planetary motion. The first law stated that the orbits of the planets were not circles, but ellipses with the Sun at one of their foci. The second stated a planet's orbital velocity is such that no matter where the planet is along its orbit, an imaginary line drawn to the Sun will sweep equal areas of the orbital plane for equal intervals of time.

The mathematical precision of Kepler's discovery failed, however, to describe the orbit of the Moon around the Earth. Despite the seeming repeatability of the Moon's monthly cycle through its phases, its path around the Earth is complicated. The Moon's orbit wobbles and precesses and its Keplerian ellipse is distorted in ways that can change the lunar month by as much as five hours.

By 1619, Kepler conceived of the third law of orbital motion that stated the squares of the planet's orbital periods are in proportion to the cubes of their distances from the Sun, completing the rules needed to mathematically predict the motions of the planets with unerring accuracy.

Kepler came close to hitting upon the theories of gravitation. He lived two generations before Newton, the eventual definer of the laws of gravitation, but Kepler made many references to his understanding of the mutual attraction between two bodies. Not only did he believe the Sun held all the planets in check with its gravitational attraction, but he believed the Earth and Moon had individual, but interacting, gravitational forces. He recognized that the Earth and Moon would have joined around a common center of mass if it were not for another undefined force keeping them separated. He also speculated that the tides were a phenomenon caused by lunar gravity. Although he defined the paths of the planets and gained lasting fame, Kepler lacked a crucial concept in physics, inertia, to complete his understanding of gravitation. It would fall

The Italian astronomer Galileo Galilei provided the first observational proof of the Copernican order of the solar system as well as ground-breaking telescopic observations of the Moon.

to his contemporary, Galileo, to discover that facet of physics and thus Kepler barely missed the discovery of gravity, the most important concept in physics.

The first known map of the Moon was drawn by the Englishman William Gilbert around 1600 using naked eye observations. Gilbert believed that the dark areas of the Moon were continents, and the light areas were seas. As was the custom in the era, Gilbert Latinized the names he applied to areas on his map. The large bright area in the center of the Moon was dubbed Mare Medilunarium. He combined present-day Mare Imbrium and Mare Serenitatis into Regio Magna Orientalis (Great Eastern Region). As an Englishman, Gilbert chose the name of Britannia for today's Mare Crisium.

The Moon Enters the Telescopic Age

Galileo Galilei is considered the first telescopic astronomer, but he was not alone in turning the telescope toward the Moon. The first telescopic lunar observations preceded Galileo's by several months when the Englishman Thomas Harriot acquired a crude telescope. On the night of July 26, 1609, 350 years to the week before humankind landed on the Moon, Harriot pointed his six-power telescope at our natural satellite and produced the first known telescopic chart of its surface. On September 12 and 13, 1610, Harriot created more detailed sketches showing 49 craters and valleys on the Moon's face. Strangely, the reality of 17th-century cartography was that Harriot's primitive Moon chart plotted lunar features more accurately than contemporary Earth maps displayed all the known terrestrial continents.

During his lengthy tenure as a professor of mathematics at the University of Padua, Galileo began corresponding with Kepler. Comparison of the two scientists reveals that Kepler ended the medieval era of science while Galileo established the modern era. In his letters to Kepler, Galileo indicated he also accepted the Copernican heliocentric theory. By May 1609, news of the telescope reached Galileo. He quickly understood its optical principles and soon produced a working model of his own. Eventually he succeeded in constructing a telescope that magnified the view 33 times.

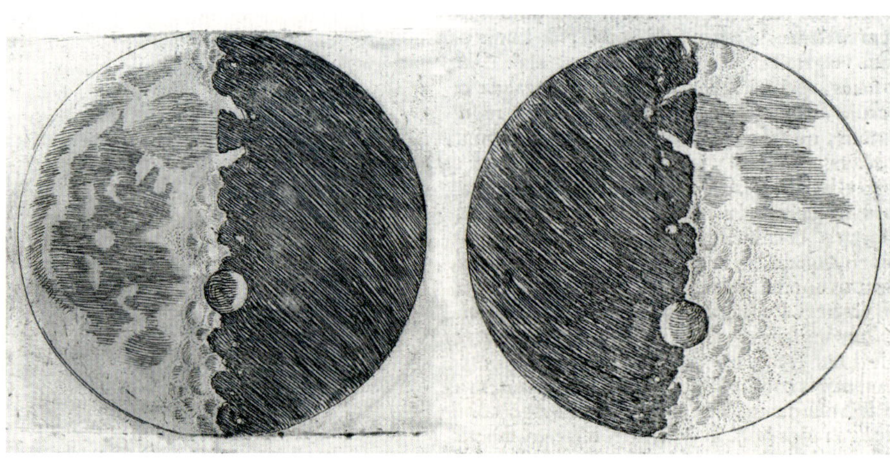

Although Galileo Galilei was the first to telescopically define the face of the Moon, he was a poor artist and his charts of the Moon published in his book Siderius Nuncius bear little resemblance to Earth's natural satellite.

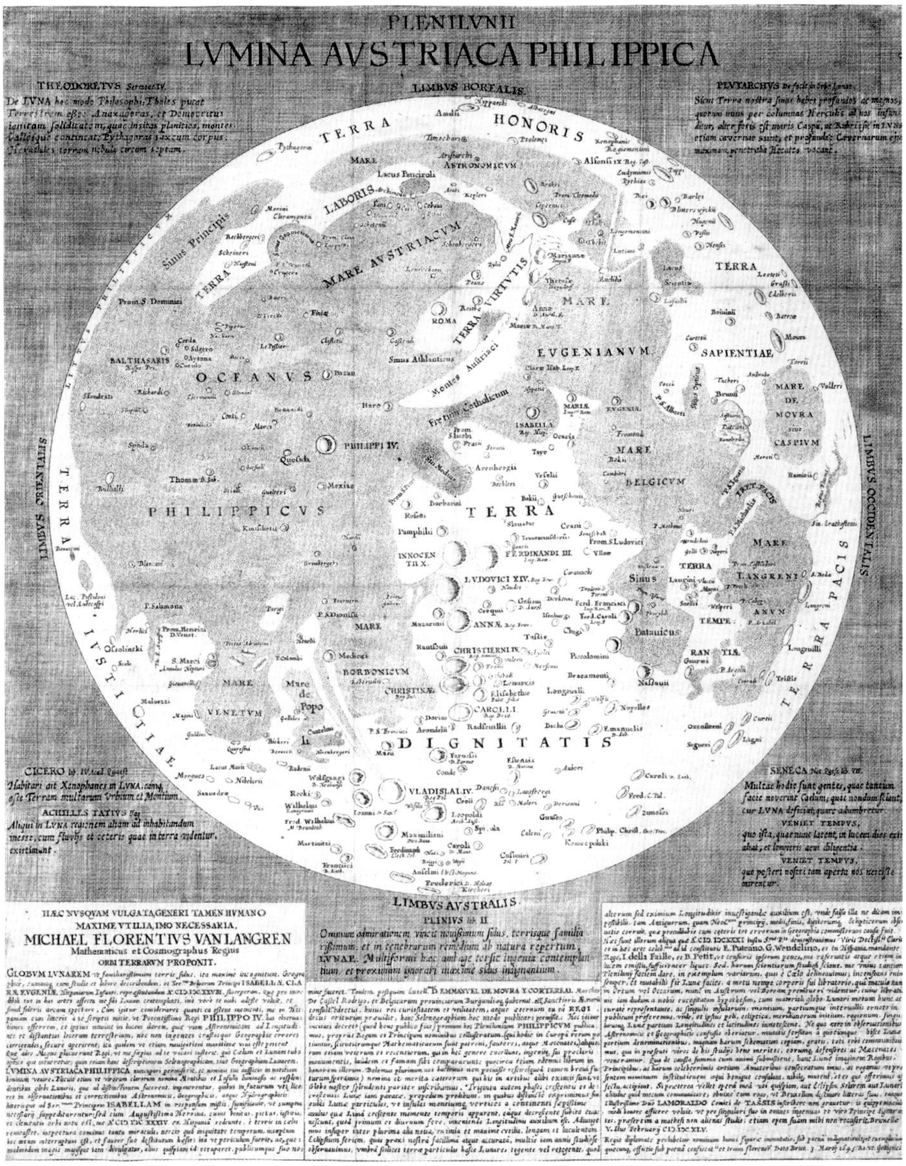

Michael van Langren, born in what is now the Netherlands, began his lunar observations in 1628 and published the first accurate and recognizable map of the Moon in 1645. Only four features named by van Langren survive on today's lunar map, including the self-named crater Langrenus.

In November of 1609, Galileo pointed his handcrafted telescope at the Moon and forever changed the science of astronomy. What Galileo saw astounded him. In just seconds, all that had been assumed about the Moon since ancient times was swept aside as he found the Moon was not the perfect Aristotelian celestial sphere that had been commonly thought for the previous 1,500 years. Instead, the Moon Galileo saw was far from smooth. Indeed, it looked pitted like corroded metal. Numerous holes covered parts of the surface, along with mountains and what were apparently valleys.

In 1628 Michael Florentius van Langren applied the family trade of mapmaking to the Moon. Known by his adopted Latinized name Langrenus, he created more than 30 sketches of the Moon at various phases, then combined all phases to show the features on the entire Moon as if illuminated by the rising Sun. Langrenus' 86-centimeter map was published in 1645 and showed 270 features.

Early 17th-century telescopes using a single light-gathering lens necessarily had extremely long focal lengths to achieve good focus. The resulting beam-mounted "aerial telescopes," such as the 8-inch telescope used by Johann Hevelius, were often over 100 feet long and required a crew of workers to aim and control the telescope.

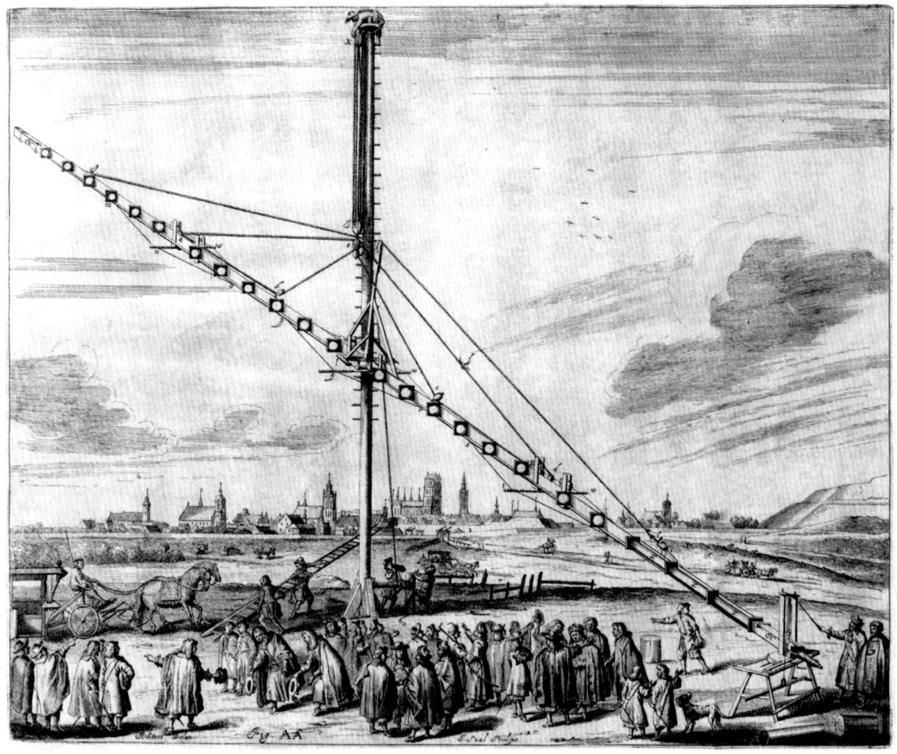

Johannes Hewelcke, known scientifically as Hevelius, built his observatory in 1630 and extensively observed and charted the Moon. In 1647, he published a lavish book about the Moon at his own expense containing his famous early Moon map and 563-page folio called *Selenographia, sive Lunae descriptio*. This was the first descriptive volume devoted totally to the Moon. In this work, Hevelius introduced the term "selenites" for his imagined lunar inhabitants. Hevelius' lunar map improved on previous works as the first to show the effects of north-south libration by representing the Moon on two overlapping circles.

With many maps of the Moon appearing by mid-17th century, the problem arose of consistent names for lunar features. The modern naming system for the face of the Moon began with a 28-centimeter map appearing in the book *Almagestum Novum* published in 1651 by the Jesuit astronomer Giovanni Battista Riccioli. His map was based on the observations of fellow Jesuit Francesco Maria Grimaldi, a mathematics professor at the University of Bologna. Riccioli chose the politically neutral theme of naming craters after astronomers and scientists. With the near universal adoption of this naming system, some historians now noted the Moon became a graveyard for astronomers.

To this day we still refer to the broad dark areas by the fanciful names Riccioli gave them after weather phenomena and states of the mind, such as Mare Imbrium (Sea of Rains), Oceanus Procellarum (Ocean of Storms), and Mare Tranquillitatis (Sea of Tranquility). Riccioli named the Moon's most prominent crater after Tycho, whose cosmological ideas were like his own. Those who professed theories different than his own, such as Copernicus and Kepler, were also honored with substantial craters, but were symbolically placed in the Ocean of Storms.

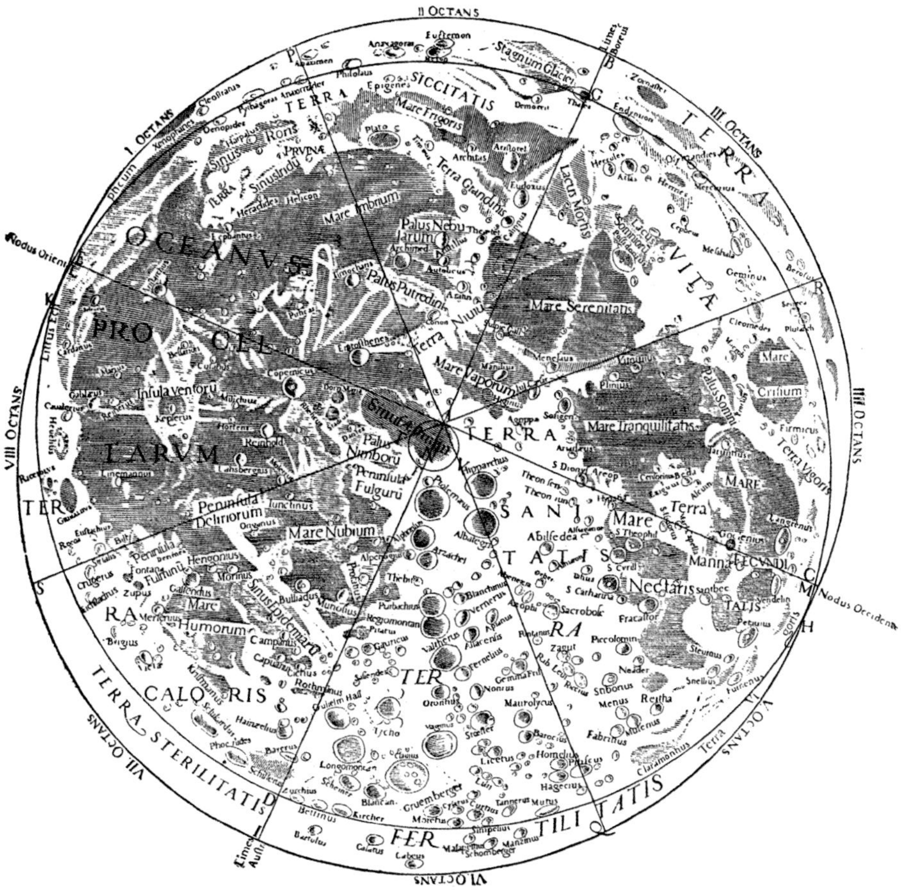

In 1651, Giovanni Riccioli published a lunar map based on the observations of his fellow Jesuit Francesco Grimaldi. The lunar feature naming scheme used by Riccioli is still used today, although Riccioli's application of names to the bright lunar highland regions is no longer recognized.

Understanding Gravity

However, 16th-century astronomers were more interested in the motion of the Moon and planets than in the bodies themselves. The paths of other worlds around the Earth and Sun were successfully described and predicted by Kepler's three laws of orbital motion, but another factor enters the planetary scheme: gravity. The full understanding of the force that attracts one body to another was not defined until the Englishman Isaac Newton tackled the problem.

A year after Newton graduated from Trinity College in Cambridge in 1665, the plague swept London, forcing the 22-year-old Isaac to flee to his mother's farm in Lincolnshire. Here, he pondered the nature of the universe and the forces that made it act in predictable ways.

It is said that once when Newton pondered what kept the Moon in the sky, he saw an apple fall from a tree. He wondered why the Moon did not also fall to Earth as had the apple. This curious thought led him down the path to the discovery of the laws of universal gravitation. His initial calculations about the force that binds the Earth and Moon started in 1665 and were not resolved for 20 years. His solutions became part of his famous work, the *Principia*, published in 1687. In this work, Newton succeeded in completing the earlier work of Galileo and Kepler and mathematically established how objects move in the universe by developing his three laws of motion and the principle of universal

Although the Englishman Issac Newton was not an observational astronomer, his late 17th century mathematical proof of the laws of gravitation became a foundation for understanding how the universe works.

gravitation. The concepts presented in Newton's *Principia* united the sciences of physics and astronomy and led to a complete understanding of the motions of celestial objects. Newton proved with mathematical precision that the only paths the planets could take were those prescribed by Kepler and confirmed by the observations of Tycho.

Robert Hooke, a British contemporary of Isaac Newton, was the first to perform experiments to determine the origin of lunar craters. He dropped musket balls into soft clay and reproduced realistic rimmed craters. In the era before asteroids were discovered, Hooke remained vexed about the source of any cosmic bullets that could impact the Moon and thus rejected impacts as a crater-making process. Hooke performed a second experiment that lent credence to the theory of volcanic origin for lunar craters, an idea that persisted until the 20th century. He let a pot of boiling alabaster slowly cool and noted when viewing the resulting pitted surface by oblique candlelight, it bore a strong resemblance to the cratered lunar surface.

The next significant lunar mapping was undertaken by Giovanni Domenico Cassini at the newly established Paris Observatory. Although Italian by birth, by the time he finished his mapping project, he was a naturalized French citizen known as Jean Dominique Cassini. The project was completed in 1679. The map was a significant improvement over previous charts but was flawed by Cassini having drawn many craters that did not exist, particularly around Copernicus.

Cassini's mapping project was a step in the evolution of charts recording ever finer details on the Moon, but as with most other lunar studies in the 17th and 18th centuries, the primary goal of the lunar work was an attempt to understand the motions of the Moon well enough to use it as an aid for ocean navigation. Cassini failed in this effort, but basic lunar science did benefit. Through his observations, Cassini developed three laws of lunar rotation:

1. The Moon rotates on its axis with a period equal to its sidereal period around the Earth.
2. The inclination of the Moon's equator to the ecliptic remains constant.
3. The ascending node of the Moon's equator on the ecliptic always coincides with the descending node of the lunar orbit on the ecliptic.

These laws were significant because they later led Lagrange, Laplace, and Poisson to mathematically investigate the dynamics of the Moon's unusual motions, leading to greater understanding of gravitation and orbital mechanics.

As the 17th century ended, the understanding of the solar system had taken a magnificent leap. The wandering planets and the Moon had been separated from superstition and mysticism and their paths in the sky could be mathematically calculated with extreme accuracy.

In the mid-18th century, Tobias Mayer became the first of two centuries of dedicated German lunar investigators. In 1748, Mayer defined the Moon's equator. Using the crater Manilius, he began a series of measurements from the crater to the eastern and western limbs of the Moon. Soon, he noted that the crater seemed to move and he could see additional details along the limb opposite the direction of movement. He soon realized the Moon wobbles as seen from Earth. This effect had been seen back to the days of Galileo's pioneering telescopic observations, but Mayer was the first to deduce the wobble was only an illusion caused by the varying geometry of the observer's viewing angle from Earth as the Moon traveled along its elliptical orbit. Mayer established the position of the Moon's axis and is credited with the discovery of lunar libration.

Most of Mayer's lunar studies centered on mastering the motion of the Moon so our satellite could be used as a navigation beacon for determining longitude at sea. For this effort, he calculated lunar tables that were accurate enough to determine longitude at sea to an accuracy of one degree and published them in 1752.

The era of gravitational astronomy, or a mathematical understanding of how objects in the solar system move and react to each other's gravity, was developed by a series of five remarkable 18th-century European mathematicians: Leonhard Euler, Alexis Clairaut, Jean d'Alembert, Joseph-Louis Lagrange and Pierre-Simon Laplace.

Leonhard Euler was the first to realize that because of the laws of gravity, a planet does not truly follow an ellipse around the Sun, and nor does the Moon follow a true ellipse around the Earth. He reasoned that as the center of mass of the combined Earth–Moon system moves along the ecliptic, the center of the Earth can be above or below the ecliptic depending on where the Moon is along its inclined orbit. Euler's observations of the Moon crossing the Sun during a total solar eclipse on July 25, 1748, erroneously convinced him the Moon had an atmosphere about 1/200th as dense as Earth's atmosphere.

Joseph Lagrange was Sardinian by birth but adopted the French version of his name. Lagrange was a mathematical prodigy and in 1780 he developed equations that described the physical librations of the Moon. Lagrange also determined the Earth's gravitational attraction on the Moon was enough to force one side to permanently face toward the Earth. He then prepared his great work *Méchanique Analytique* where he continued the work of Newton and refined the laws of universal gravitation. Today, Lagrange's name is most associated with the "Lagrange points," although three of the five Lagrange points were discovered by Euler. These five points are orbital positions relative to a large body where a satellite will remain in a stable fixed position rather than in an orbit that constantly changes position relative to the parent body.

Pierre-Simon Laplace gained fame by solving the problem of the acceleration of the Moon that was discovered by Halley. Laplace concluded the acceleration of the Moon is due to the secular variation of the eccentricity of Earth's orbit around the Sun. His subsequent astronomical works, a five-volume set published between 1798 and 1827, popularly known as *Méchanique Céleste*, are regarded as the sequel to Newton's *Principia*.

The evolution of celestial mechanics allowed more than just positional predictions of the Moon and planets. Using the mathematical tools provided by Kepler, Newton, and Laplace, it was now possible to calculate the mass of the Moon and planets, adding another datapoint to our understanding of these worlds.

Astronomical Dynasties Established

William Herschel, a German émigré who as an Englishman became the "father of stellar astronomy," also paid attention to the Moon. Between 1777 and 1779, he measured the height of several features on the Moon and found the lunar elevations recorded by Hevelius were excessive. He found most lunar heights were under two kilometers and only a few exceeded two kilometers.

In 1787 Herschel electrified the young science of selenography with his report of ongoing volcanic eruptions on the Moon. But on this point, Herschel was wrong. The most widely accepted explanation is that he saw the unusually bright crater Aristarchus under the illumination of earthshine. Its extreme reflectivity makes it prominent, even in feeble earthshine. This illusion led Herschel astray and the resulting controversy over lunar volcanism erased his interest in the Moon, as he never again turned his telescope to that celestial target.

The fashionable English portrait artist John Russell used a telescope constructed by William Herschel to undertake a mapping project of the Moon. Starting in 1785 and continuing for 18 years he made about 200 pencil sketches of the Moon. He also produced a large pastel drawing of the Moon in 1795 that would pass today as an excellent eyepiece rendering of a waxing gibbous Moon. In 1805, Russell published a 38-centimeter map of the Moon and produced the first commercially available Moon globe. In modern times, six copies of Russell's magnificent lunar globe still exist. The 20-centimeter globes showing only the Earth-facing lunar hemisphere were individually engraved, hand painted, varnished, and mounted in an elaborate single-stem brass stand that allowed the globe to reproduce the effect of lunar libration.

A contemporary of Herschel's was Johann Hieronymus Schröter. If Herschel was the father of stellar astronomy, then fellow German-born astronomer Schröter was certainly the father of selenography. In 1781, Schröter applied for the position of chief magistrate of the small town of Lilienthal, near Bremen, where his light governmental duties allowed plenty of time for astronomy. Over the next 30 years, Schröter used a considerable portion of his large inherited estate to build an extensive observatory complex in Lilienthal from which he continuously observed the Moon and planets.

In the late 18th century, the German astronomer Johann Schröter performed the first systematic long-term observations of the Moon, specifically looking for changes in the lunar surface.

In the winter of 1787, Schröter's attention was drawn to the Moon by Herschel's reports of lunar volcanism. For the next two years, Schröter devoted himself to creating a 1.2-meter map of the Moon, but soon realized this was too ambitious and scaled his plan back to producing charts of selected regions. The objective that guided Schröter in his lunar work was the search for changes on the Moon's surface.

The Lilienthal astronomer was the first to pay serious attention to lunar rilles. Schröter did not discover rilles, but he was the first to describe the trench-like features by that designation and he charted a dozen of them. Over the course of his lunar work, Schröter added 70 new names to Riccioli's list of features.

Schröter broke new ground by studying the limb, measuring heights, investigating rilles, and discovering that the maria were not flat, but were marked by ridges. Schröter's failing that led to his discredit was his belief in a considerable lunar atmosphere. Schröter also concluded that lunar craters were volcanic. He created scale models of craters from sand, where he measured the material in the walls compared to the volume of the crater pit. He found the rim material would fill the interior pit and thus came up with Schröter's Rule that dictated the crater's rim material equaled the amount of material excavated from the crater's interior, thus craters would form with little loss of material from the Moon.

The Bavarian physician-turned-astronomer Franz von Paula Gruithuisen had a deep appreciation for the lunar work of Schröter, but he is known primarily for his unusual beliefs about the Moon. With his 100-millimeter refractor, he spotted five craterlets on the floor of Plato and pondered why they had not been seen before. Gruithuisen studied lunar rilles and decided they were either dry riverbeds or roads built by lunar inhabitants. Gruithuisen also agreed with Schröter that the Moon possessed vegetation and an atmosphere.

Gruithuisen became director of Munich Observatory in 1826 and remains noteworthy for being among the first to propose a meteoric origin for craters. However, his lunar work was clouded by a pro-inhabited Moon prejudice that was even stronger than Schröter's. He professed that some of the clefts on the Moon were "artificial clearings in the forests, answering the purpose of roads."

After being ridiculed about his assumption that some lunar features were artificially created, he turned his attention to the formation of craters. In this area, Gruithuisen was ultimately proven correct in his idea that craters were formed by the bombardment of the Moon by meteorites and comets early in its history. In the 1830s, Gruithuisen received support for his theory of meteoric origin for craters from mining engineer Carl Ludwig Althans. Repeating Robert Hooke's 17th-century crater-forming experiments, Althans dropped iron balls from a height of eight meters into mixtures of plaster of Paris and concrete. The resulting impacts reproduced lunar-like craters, leading Althans to conclude lunar craters were of meteoric origin.

John Herschel, William Herschel's son, followed in his father's celestial footsteps and investigated all matters of cosmic wonders, including the Moon. The younger Herschel's name is forever connected to what is now called the "Great Moon Hoax." On August 25, 1835, a series of six articles appeared over a week's time in the *New York Sun*. These reported the supposed discovery of a civilization on the Moon by Sir John Herschel, then the preeminent astronomer in the world. The articles reported that this supposed discovery was made with a fantastically huge new telescope located at the Cape of Good Hope in South Africa. At the time the articles circulated, John Herschel himself was indeed at the Cape of Good Hope conducting his now famous astronomical surveys from that location. The logistics of the day prevented word of the New York articles detailing his supposed discovery from reaching him in a timely manner to allow him to refute the claims.

The series of articles were initially a public sensation, and the circulation of the *New York Sun* skyrocketed. Succeeding installments of the amazing story went on to describe the newfound life on the Moon in impossibly minute detail, a factor that eventually led to its unmasking. The true purpose of the hoax seems to have been a ploy to boost circulation of the newspaper.

The German astronomer Wilhelm Lohrmann used scientific measurements to precisely plot lunar features to produce one of the finest lunar maps of the 19th century. Unfortunately, bureaucracy delayed the publication of the maps for decades after his death in 1840.

The Evolution of Systematic Lunar Studies

At the age of 25, Wilhelm Gotthelf Lohrmann decided to improve the lunar map by charting lunar features using scientific methods of measurement. In 1822 Lohrmann erected an observatory on the upper floor of his house in Pirna, a suburb of Dresden. The prime instrument was a 12-centimeter Fraunhofer achromat with a filar micrometer. Using these instruments, he measured 79 craters on the Moon with great precision and used these anchor points to make 25 ink drawings to chart the Moon on a scale of 97.5 centimeters. The first four of the 25 maps were published in a 1824 volume and included 130 pages rich in the history of lunar studies, his method of charting the lunar regions, and detailed descriptions of lunar features. Lohrmann finally completed all 25 charts in 1836. The engraving process to print the charts progressed slowly, but he published a detailed 38-centimeter map in 1838. Unfortunately, he succumbed quickly during an 1840 typhoid epidemic and never saw his complete works published.

Eleven years after Lohrmann's death, the publication of the final charts was undertaken by Julius Schmidt, himself then a well-known selenologist. However, Schmidt was a poor businessman and Lohrmann's beautiful lunar charts did not see print until 1878, 38 years after their creator had died. Once all 25 of Lohrmann's charts had been publicly released, it was clear that his early 19th-century efforts surpassed everything else previously attempted. A total of 7,178 craters and 99 rilles were charted. However, the decades long delay in publishing his complete works hid from his contemporaries that Lohrmann was the first of the modern lunar cartographers.

The German astronomers Wilhelm M. Beer and Johann Heinrich von Mädler became an important force in planetary astronomy in the early part of the 19th century. They published the first large-scale scientific map of the Moon, and the two astronomers' names are forever combined in the history of lunar study.

The pair began their association in 1828 when Beer was introduced to Mädler by Alexander von Humboldt, the German explorer and naturalist. They joined forces in cooperative astronomical studies using the 95-millimeter Fraunhofer achromatic refractor in Beer's rooftop observatory located in the Berlin Tiergarten, an area near today's famous Brandenburg Gate.

When it became apparent that Lohrmann was unable to complete work on his lunar map, Beer and Mädler decided to create one of their own. Beginning in April 1830, the two observers spent 600 nights charting the Moon at a magnification of 300. Using micrometer measurements of 106 primary points, Mädler created a gridwork by which to precisely position all other features on the Moon's disk. The heights of 800 individual lunar peaks were determined, and the diameters of 150 craters were measured. Their one-meter-diameter chart was published as four sheets between 1834 and 1836 and added 145 new feature names to the Moon. To honor their mutual friend who had introduced them eight years earlier, the two observers named a vast dark area on the Moon's eastern limb Mare Humboldtianum. Their work remained the best lunar map for the next 40 years.

William Radcliffe Birt recognized the enormous difficulty of detailed manual lunar cartography. Through his initiative in 1864, the association for the Advancement of Science appointed a "Lunar Committee for Mapping the Surface of the Moon" on a chart 2.5 meters in diameter. The Moon was split into hundreds of subdivisions to divide the work among many capable selenographers. Before significant work could be done, the British Association had withdrawn its support, and the committee was disbanded in 1870. Birt persevered and in 1878 he formed, along with Thomas Webb and Edmund Neison, the Selenographical Society. However, this organization also fell short of achieving significant lunar research, primarily because of Birt's death in 1881 and Neison's departure for South Africa in 1882.

In an 1839 speech to the Chamber of Deputies in Paris, the astronomer François Arago announced the invention of the photographic daguerreotype. Louis Daguerre himself was too ill to address the assembly, but he had brought photography to a commercially viable state early in the 19th century. Daguerre tried to photograph the Moon in 1839. However, his plates had a photographic speed some 30,000 times slower than today's medium speed photographic processes and the smear he recorded bore no resemblance to the Moon.

Many of the first astronomical photographs taken during the 19th century were obtained not by professional astronomers, but rather by amateurs knowledgeable about the photographic process. Less than a year after Daguerre had invented his process, the American John William Draper used the method to photograph the Moon on March 23, 1840. His exposure required 20 minutes and produced an image 25 millimeters in diameter. By 1850, the photographic process had advanced to the point where Draper managed to photograph the Moon with an exposure of only several seconds. Also in December 1850, the Boston daguerreotypist John Adams Whipple, assisting William Cranch Bond, the director of Harvard College Observatory, succeeded in imaging the Moon with the 38-centimeter Merz refractor.

Using the more sensitive wet-plate photographic process perfected in 1850, the Englishman Warren De la Rue succeeded in 1857 in taking amazing stereoscopic photographs of the Moon from his private observatory with a 33-centimeter reflector. The Moon's image spanned 28 millimeters on some of De la Rue's negatives and could be enlarged 20 times.

In 1864, the New York lawyer turned astronomer Lewis Morris Rutherfurd also succeeded in creating stereoscopic images of our natural satellite. These images were popular parlor entertainment when viewed through hand-held stereoscopes popular at the time.

The keeper of the lunar flame during the late 19th century was Julius Schmidt. Schmidt's interest in the Moon was sparked when he was a teenager using a small refractor with lenses ground by his father. He began to make drawings of the Moon and realized his works could have scientific value if he could access a more powerful instrument. In 1841, Schmidt moved to Hamburg, where he was able to use the telescopes at the local observatory. In 1845, Schmidts went to Bonn Observatory where he worked as an assistant for eight years. Here he continued his lunar work and had occasional use of the 23-centimeter refractor at Berlin's Royal Observatory. Schmidt noted that on May 17, 1853, the instrument revealed 78 craters on the floor of Clavius.

In 1851, Julius Schmidt was approached by publishers to complete the editing of Lohrmann's earlier maps, but he did not complete work on them until 1878. In the mid-1850s Schmidt was invited to take charge of a private observatory at Olmütz in Moravia. Here, he had access to a Fraunhofer refractor equipped with a filar micrometer. He quickly put this equipment to use and over the next three years he made 2,894 height measurements of lunar mountains. In 1856, he also published *Der Mond*, a 164-page treatise that was probably the first discussion about the Moon by someone familiar with geology.

In 1858, Schmidt assumed the directorship of Athens Observatory in Greece, a position he would hold for the rest of his life. The 16-centimeter Plossl refractor became his primary lunar instrument for the next 25 years. In 1865, Schmidt decided to compile his fragmentary observations into a general topographic map of the Moon consisting of 25 sections, each 38 by 38 centimeters, showing the Moon as a disk 1.8 meters in diameter. Some of the individual sections were based on 150 separate drawings. Schmidt was fascinated by rilles and by 1866 he had cataloged 425 of these features in his monograph *Concerning Rilles on the Moon.*

By 1874, Schmidt's great chart was finished and surpassed anything previously done with the Moon. Lohrmann had plotted 7,177 craters and 77 rilles earlier in the century. Beer and Mädler later recorded 7,735 craters and 71 rilles. Schmidt's lunar chart showed 32,856 craters and 348 rilles. Schmidt's great map was a tremendous achievement for a single observer, but it did have the flaw of being inconsistent because it had taken him 32 years to complete the work using 13 separate telescopes. Although Schmidt worked in Greece, the Prussian Government published his work in 1878 in Berlin as *Chart of the Mountains of the Moon.*

The Evolution of the Pre-Space-Age Moon

The Scotsman James Nasmyth was a 19th-century amateur astronomer and self-taught master telescope maker whose lunar observations profoundly influenced how the surface of the Moon was perceived over the next three-quarters of a century. At the age of 19, Nasmyth engineered an improved method for casting speculum telescope mirrors.

Nasmyth's skill at casting and polishing telescope mirrors gained him quick fame as a telescope maker and he set his sights on a half-meter telescope. He mounted the telescope tube on trunnions like a cannon and modified the Cassegrain optical system by placing an additional angled flat in the optical path that routed the telescope's image through one of the hollow trunnions. This design allowed the operator to view through a fixed eyepiece while remaining seated on the side of the telescope mount, no matter where the telescope pointed in the sky. This optical design still bears Nasmyth's name today.

Systematic lunar observations began with the half-meter telescope in 1842. Nasmyth observed at every opportunity and concluded that most of the Moon's surface features were created by volcanic action. He reasoned that the large size of many lunar craters, compared to earthly volcanic craters, was due to the reduced lunar gravity. Nasmyth's conclusions about the volcanic origin of lunar features carried a great deal of weight well into the 20th century.

In 1874, Nasmyth and James Carpenter, a former assistant at Greenwich Observatory, co-authored the book *The Moon, Considered as a Planet, a World, and a Satellite*. What separated Nasmyth's and Carpenter's book from the numerous previous lunar volumes was theirs was the first attempt to explain why the Moon looks like it does. Their work created a new variation of lunar science, selenology, or the study of lunar geology.

The era of reliance on the artistic skill of the astronomer to record and map the Moon came to a merciful end at the close of the 19th century. For the previous half-century attempts to photograph the Moon with the daguerreotype and wet plate processes met with limited success and were considered scientific curiosities instead of hard science.

The usefulness of photography to chart the Moon advanced dramatically with the perfection of the dry plate process in 1871. Richard L. Maddox substituted gelatin for collodion and used silver bromide as the light-sensitive agent. Maddox's new plates could be used dry, prepared long in advance, and developed at leisure. The new dry plate was 200 times faster than the daguerreotype and a new era in astronomy was about to begin. The image scale of the long focal length professional refractors at the end of the 19th century quickly yielded photographs that reached the limits of what could be resolved with photography.

Starting in 1894, Moritz Loewy and Pierre Puiseux used the telescope affectionately called the *Grande Lunette* at Paris Observatory to take 2,000 images of the Moon. Their joint 1896 work titled *Atlas Photographique de la Lune* comprised 77 sheets at scales up to 1:1,500,000. It was one of the first European lunar photographic atlases and was regarded as the finest available until after World War II.

In 1885, Julius Franz began his lunar work using the heliometer at Königsburg Observatory to make precision measurements of the Moon. Franz combined his heliometer measurements with photographic measurements to establish 150 selenographic positions and the heights of 55 craters. He used this information to create the first contour map of the Moon. Franz also noted the "continents," now known as the highlands, on the southeastern quadrant of the Moon were higher than those on the northwestern quadrant. Franz also embarked on a mission to name many of the Moon's smaller dark spots as maria, or lunar seas. Some of his designations, such as Mare Spumans and Mare Undarum, survive today.

At the end of the 19th century, astronomers continued to debate whether lunar craters were created by volcanism or meteorite impacts. The United States Geological Survey (USGS) was founded in 1879, and within two years, this fledgling Government agency was already exploring meteorite craters and casting an eye on the Moon.

In 1874, James Nasmyth, along with James Carpenter, published the book *The Moon, Considered as a Planet, a World, and a Satellite*. So influential was the book that for the next 50 years it shaped the erroneous notion of volcanic origin for lunar craters.

In 1892, Grove Karl Gilbert, the first head of the U.S. Geological Survey, used the 66-centimeter refractor at the U.S. Naval Observatory in Washington, DC, to study the Moon. Gilbert was the first professional geologist to determine the Moon's major features to be created by meteorite and asteroid impacts.

A Geologist Considers the Moon

The key to cracking the secret of the origin of the Moon's features had its genesis in 1891 with Grove Karl Gilbert, the Chief Geologist of the USGS. Already fascinated by the Moon, he became attracted to reports of large amounts of meteoric iron around Coon Mountain in the Arizona Territory. The geologic feature was called a mountain in name only as a climb up the slope revealed a gigantic hole in the ground nearly 1.5 kilometers across and sunken 200 meters below the surrounding desert floor. Today, this area is known as Meteor Crater. Gilbert speculated the crater might be a "scar produced on the earth by the collision of a star." As a geologist, Gilbert advocated for an impact as an origin for the crater's formation, but the hitch in his theory was the lack of other craters on Earth. It was inconceivable that the Moon could be peppered with cosmic buckshot while the Earth remained unscathed.

Traveling to Arizona, Gilbert applied the science of the day to the Coon Mountain crater. He calculated the excavated volume of the giant hole to be 63 million cubic meters, which was identical to the volume of ejecta piled up around the crater, strongly implying the mass of debris came from the crater.

The 19th-century intuition about the physics of meteorite impacts led to the expectation that the remains of a giant meteorite would be lodged in the ground like a bullet stuck in tree bark. However, Gilbert's magnetic survey showed no deviation between readings on the crater's rim and the center. Gilbert was reluctantly forced to ignore his instincts that the crater was of meteoric origin and concluded the crater was a maar, or crater created by a volcanic steam explosion, and the presence of meteorites in the area was a coincidence.

His near miss with crater reality did not deter Gilbert's interest in lunar craters. For 18 nights in August, September, and October 1892, he observed the Moon with the 66-centimeter refractor at the Naval Observatory in Washington, DC. His resulting paper, *The Moon's Face: A Study of the Origin of Its Features*, was presented as his retiring address as president of the Philosophical Society of Washington. Gilbert was not the first to suggest that an impact was the source of lunar craters, but his paper was the first to apply scientific arguments that all craters, from the smallest to the largest, were the result of meteorite impacts.

Gilbert also noted the system of parallel ridges on the Moon that he termed "sculpture." When they were plotted on a chart, "they converge toward a point near the middle of the plain called Mare Imbrium, although none of them enter that plain." He concluded flying debris heaved from a colossal explosion in the Imbrium area gouged the ridges and furrows. Being the first to recognize that Imbrium was the site of a gigantic impact, he was able to date features on the Moon as having occurred before or after the Imbrium impact. Thus, Gilbert was the first to apply the science of stratigraphy to the Moon, or the system for classifying the relative age of various features by noting how younger features overlay older features. He correctly assumed the heavily cratered southern highlands "probably represent the general condition of the surface prior to the Imbrium event." Another of his observations was that craters smaller than 20 kilometers were simple bowl-shaped excavations, those larger than 20 kilometers had central peaks, and those larger than 150 kilometers rarely had central peaks. He also observed that terraces within the larger craters were formed by landslip and the floors of craters were almost always below the plain of the surrounding crater.

Overall, Gilbert's conclusions about the Moon and how its face evolved were advanced for the times. Unfortunately, his papers about the Moon and its evolution were lost in an obscure institutional archive for many decades and his deductions remained unrecognized until the mid-20th century.

In 1946, Harvard geology professor Reginald Aldworth Daly lent his support for the impact theory and allied himself with Gilbert's views. In the same year, Robert Sinclair Dietz also cited Gilbert's work and added some of his own ideas, arguing in favor of all lunar craters being of impact origin. Dietz listed eight properties of lunar craters that he felt segregated them from terrestrial volcanic craters. In particular, he attacked two features held dear by endogenists. First was the circularity and radial symmetry of craters. Dietz pointed out it was volcanic craters that were asymmetrical, not the other way around. The other point was the issue of craterlets on the central peaks of craters. These were seen as the smoking gun indicating the peaks were volcanic. Dietz countered they were instead shadow effects and not volcanic pits. Forty years later, Lunar Orbiter satellite photographs would prove him correct.

To prove his point about lunar impacts, Dietz searched for similar impacts on Earth. He eventually found that the Sudbury Structure in northern Ontario is an ancient impact feature. His proof was the discovery of shatter cones: grooved structures in rock that can only be caused by titanic shock forces created by either a nuclear explosion or an asteroid impact. With this, scientific opinion further swayed in favor of impacts creating lunar craters.

Walter Goodacre, the second Director of the British Astronomical Association's Lunar Section, produced what was considered the first "modern" map of the Moon in 1910. Goodacre's map was drawn from his observations through a 20-centimeter Cooke refractor and photographs taken at the Paris Observatory and Mt. Wilson. For positional control of lunar features, he used 1,433 points measured by fellow Englishman Samuel A. Saunder. In 1931 his original map was reproduced on a scale of 76 centimeters in his book, *The Moon*.

In the first half of the 20th century, a succession of astronomers and geologists produced a series of insightful works that ring true to modern lunar facts, but their works were lost in obscurity. In 1916, the Estonian astronomer Ernst Öpik was the first to produce studies showing that the mass distribution of meteors, asteroids, and comets was the same as the size distribution of craters on the Moon. This was a direct link pointing to impact origin for craters. Öpik's work was lost in the fog of World War I. Written in Estonian and Russian, and published in obscure journals, his work would not be found by the western world until the 1950s.

In 1949, Ralph Baldwin published the *The Face of the Moon*, the first book to "get it right" about the impact origin and geologic nature of lunar features. Baldwin's book influenced scientific events in the 1950s that eventually placed the American space program on course to explore the Moon.

Finally Getting the Moon "Right"

The glacial pace of lunar investigations in the 1930s and 1940s, necessarily slowed by World War II, soon received a far-reaching stimulus in the form of Ralph Belknap Baldwin. Although he was educated as an astrophysicist, he was earning a living in the family business, Oliver Machinery Company, in Grand Rapids, Michigan. Baldwin has been described as having an almost Newton-like mind in that he could focus with great intensity on a subject.

Baldwin was steered toward the Moon in 1941 while waiting for a lecture at the Adler Planetarium in Chicago. There he noticed the photographic transparencies of the Moon on public display. His attention was arrested by the grooves on the Moon that 50 years earlier Gilbert had described as "sculpture." He did not know about Gilbert's analysis half a century before, so he figured out the nature of the sculpture on his own.

In 1942, Baldwin wrote two papers in which he presented his conclusions about the Moon. He submitted them to the major astronomical and astrophysical journals, but they were rejected, as at the time the editors did not consider the Moon to be a serious astronomical subject. This only strengthened his resolve to study the Moon and Baldwin found he had the area of lunar research almost entirely to himself. History shows he made good use of his scientific solitude.

Ralph Baldwin humbly entered the world of lunar research by publishing his works in the magazine *Popular Astronomy*. In the first article, Baldwin concluded that Mare Imbrium was too big to be volcanic and was created by a huge explosion. The grooves across the Moon he had noted "were caused by material ejected radially from the point of explosion." Baldwin's second article described how he envisioned the impacting body flattening out by shock and excavating a cavity laterally. This was a significant leap in contemporary thinking about the impact process of lunar crater and basin formation. Baldwin remained focused on the Moon for the next several years, but his attention was divided between the Moon and his war contribution to the development of the artillery proximity fuse.

Through this time, Baldwin prepared a book-length manuscript detailing the results of his own lunar observations and research. In this solitary work exploring all that was available about the Moon at the time, Baldwin acted as astronomer, geologist, geophysicist, geochemist, and physicist while creating the most far-reaching and influential lunar book ever written, *The Face of the Moon*, published by the University of Chicago in 1949. It has been described as the first book that "got it right" about the Moon. Baldwin's lunar conclusions would grab the attention of some of the foremost scientists in astronomy and geochemistry and put them, and eventually the United States, on a collision course with the Moon.

An important part of Baldwin's argument for the impact origin of craters was using meteor observation data to establish the size-frequency distribution of meteors. Like Öpik's lost data from the World War I era, Baldwin also noted the curve of that function perfectly matched the size-frequency distribution of craters on the Moon. He also thought the large mare areas were lava flows. Most importantly, Baldwin recognized that the large circular mare areas lie within huge craters. *The Face of the Moon* was a post-war tipping point that marked the acceptance that impacts created the face of the Moon.

As with many important scientific books, Baldwin's volume did not sell very well, but it did reach some important readers. One of the most important was Harold Clayton Urey, the Nobel-winning chemist from the University of Chicago. After reading Baldwin's book, Urey became engrossed with the Moon. His interest was driven not by the desire to explain its features but by his expectation that the Moon was a piece of the primordial solar system and much older than the Earth itself. This started a chain of events that lunar geologist Don Wilhelms describes as leading to the choice of the Moon as America's goal in space in the 1960s.

Urey's interpretations about the Moon carried a lot of weight because of his scientific status and his enthusiasm. This was at a time when most astronomers still felt the Moon was an uninteresting, cold, dead world. While Urey was right about a lot of things regarding the Moon, over the next two decades he would find some of his fundamental expectations about the Moon were flawed, including the idea that it is an unchanged time capsule preserving a piece of the early solar system.

For 350 years after the perfection of the astronomical telescope the study of the Moon was, for the most part, the solitary pastime of only a handful of people. At any one time it was rare that more than one or two people worldwide were engaged in serious lunar research. The torch was unknowingly passed throughout the centuries of lunar studies.

After those centuries of individual selenology, much duplication and name confusion were created by multiple generations of maps on which the naming of lunar features remained the whim of the individual observer. Well into the 20th century there was no accepted lunar naming authority.

Mary Adela Blagg and Dr. Karl Müller undertook the task of sorting everything into a comprehensive uniform format. The specific designations for 672 named formations were established in 1913. A 14-section map was drawn based on the measurements of Franz and Saunder to reference the named formations. Commission 16 of the International Astronomical Union formally adopted the lunar names in 1935. In 1938 the British Astronomical Association issued *Who's Who In the Moon*, a valuable resource for locking in lunar name designations and providing biographical information about the individuals memorialized by named lunar craters.

Today, a lunar chart is dotted with features named after great thinkers and a map of the Moon reads like a catalog of great scientists and philosophers that pondered and studied the universe since the age of scientific reason began over 2,000 years ago. Thanks to those immortalized on the face of the Moon, we have discovered not just the Moon, but also how the cosmos works.

Any history of astronomy is incomplete without an accounting of the individuals who defined the Moon and were principal players in the advancement of science. The succession of British lunar experts from the late 19th to the mid-20th century reads as a lineage, one passing the torch to another as they made their contributions to lunar science. Indeed, the story of these lunar pioneers and the growing cadre of international lunar specialists deserves its own historical volume. But the slow solitary pace of lunar progress would soon change with the rising of a massive rocket from the steppes of Kazakhstan on October 4, 1957. The appearance of Sputnik in the October sky changed the world forever.

2

The Moon Enters the Space Age

The launch of Sputnik in October 1957 split lunar exploration in two directions. One track remained the slow-paced academic study by luminaries like Ralph Baldwin and Harold Urey and newcomers like Gerard Kuiper and Eugene Shoemaker, who would soon become lunar household names. The other track was the politically driven, fast-paced response by America and Western Europe to the challenge Sputnik posed: Soviet domination of space.

At this point the almost royalty-like lineage of lunar observers was splintered into many branches, and the number of lunar astronomers, geologists, and chemists grew exponentially. During the early Space Age, advanced amateurs like Britain's Patrick Moore and America's Alika Herring became the public voice of the Moon as they spread their lunar insights to a public fascinated by the accelerating race to the Moon. The exciting glamor of gleaming Moon rockets soon pulled the public's eye away from isolated mountaintop observatories. What was science fiction the year before was now a reality, and spaceships destined for the Moon were on engineers' drawing boards.

The history of lunar exploration shows that it was Cold War politics that pushed lunar exploration into a rocket-powered frenzy in the mid-20th century. As missile technology matured in the decade after World War II, both American and Soviet technicians raced to be the first to launch the first artificial Earth satellite. When the USSR shocked the West by launching Sputnik, national prestige, set on the stage of capitalist versus communist politics, was the primary driving factor in the sudden rush to the Moon. From this point on, the study of the Moon ceased to be the passion of a select few. Now vast technical armies were marshaled in both the United States and Russia in the effort to be the first to put a man on the Moon.

Before NASA existed, two key players in the American response to Sputnik were the U.S. Air Force and the U.S. Army. Both mounted technically challenging and daring lunar flight programs. However, fate reserved the first lunar success for the Soviet Union. Their Luna 1 probe escaped Earth's gravity a mere 15 months after the dawn of the Space Age. Although Luna 1 missed the Moon, its flight escalated the political and scientific stakes between the superpowers.

Facing page: Gerard Kuiper and Harold Urey were leading lunar authorities in the 1950s, but had opposing views on the physical nature of the Moon, with Kuiper championing the "hot" formation of the Moon while Urey favored the "cold" formation of the Moon.

Harold Urey

Gerard Kuiper

For the post-World War II generation, the concepts of spaceflight and exploring the Moon were indivisible from harsh Cold War politics and intense propaganda between America and the Soviet Union. The Moon was as much a political as well as scientific target. People born after the Apollo era, and particularly those who have come of age in the 21st Century, have no concept of the politically driven push to "beat the Russians" to the Moon and why we felt compelled to do so.

Any modern story about the Moon is incomplete without a look at Space Age politics as well as the technical struggles involved in reaching our nearest celestial neighbor. Drifting into the recesses of history are the dozens of unmanned robotic Moon probes launched in the decade before Apollo's epic flights and the ongoing efforts to explore the Moon in the early 21st century. It is important to preserve the memory of the hard-fought technical struggle to explore the Moon in preparation for Apollo and the continued 21st-century return to the Moon.

Mapping the Moon from Earth

Before rockets could send probes to the Moon, academics needed to define the Moon's surface so engineers could create spacecraft that could safely land and explore. There were few lunar scientists in the 1950s and contemporaries of lunar champion Harold Urey could be counted on the fingers of one hand. But when the Moon became a goal of national policy in the 1960s, there was an explosion of new lunar and planetary specialists, many concentrated in the geological field. Urey would often argue with geologists and those whom he would call second-rate scientists because they were not "selenologists." Urey would buck many of the accepted ideas about the Moon until the bitter end, when Apollo astronauts brought home the undeniable truth about the history, composition, and nature of volcanism on the Moon. However, to his credit, once his own theories were proven wrong, Urey accepted reality and no longer held any grudge against those who scientifically opposed him.

Gerard Peter Kuiper was another highly influential lunar scientist in the 1950s and 1960s. A Dutch immigrant, Kuiper became chairman of the University of Chicago astronomy department and director of the then combined Yerkes and McDonald Observatories from 1946 to 1949 and again from 1957 to 1960. When Baldwin's *The Face of the Moon* was being prepared for publication in 1948, it was Kuiper whom the publisher asked to referee the book. At the time, Kuiper was almost single-handedly carrying the science of planetary astronomy, and he had recently discovered the atmosphere surrounding Titan, Saturn's largest moon. But at the time Kuiper felt he knew too little about our own Moon to effectively referee Baldwin's book. The task was taken on by Fred Whipple at Harvard College Observatory.

In 1953 Kuiper's interest in the Moon locked into high gear because so little was agreed upon regarding its origin and features. Like Urey, Kuiper also saw the Moon as a geologic record of the early solar system. He even did something other astronomers were loath to do because the study of the Moon

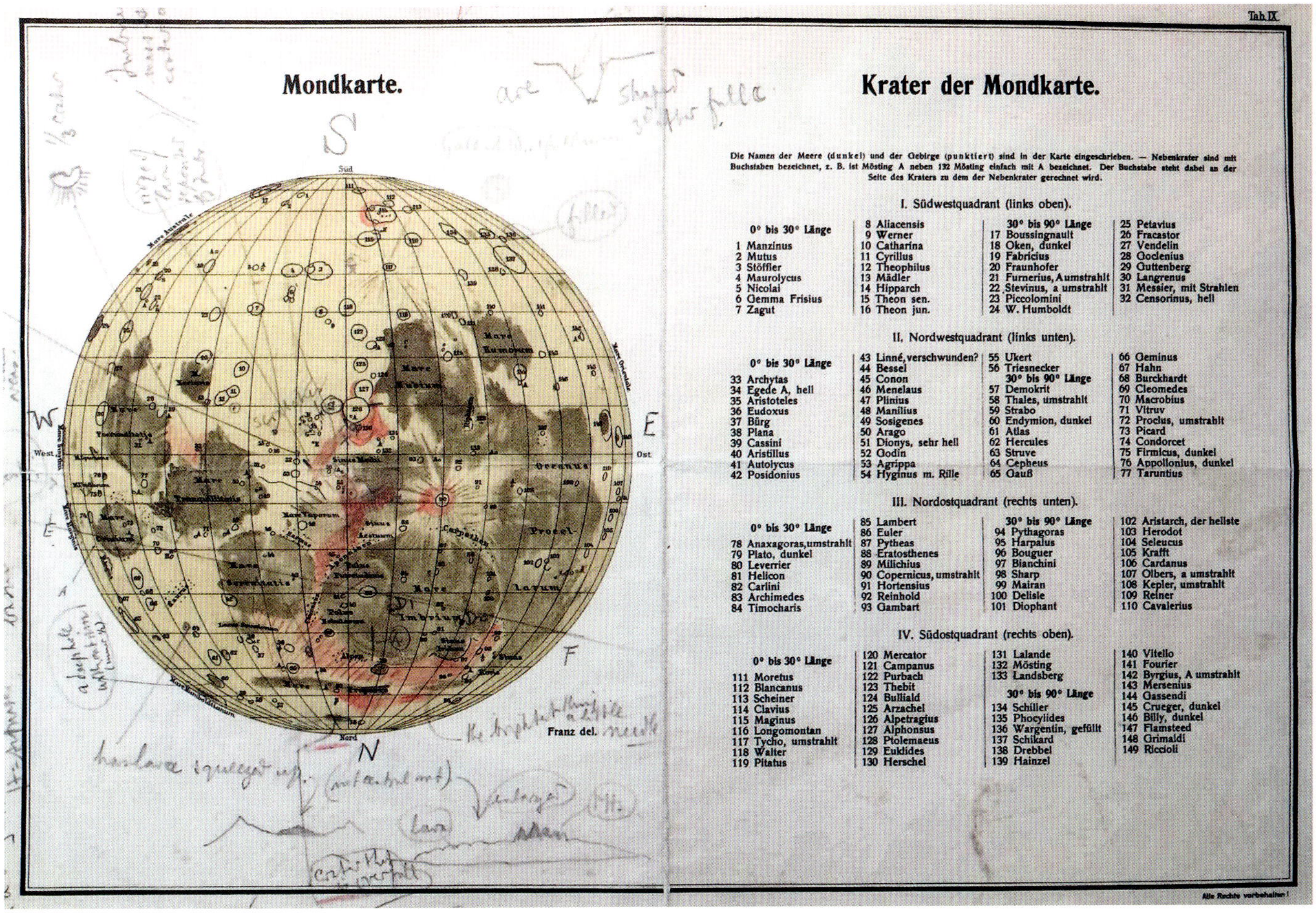

was considered a dead end in astronomy, he observed the Moon through a telescope. On nights when the 2.1-meter McDonald Observatory reflector, then the third largest in the world, was out of stellar service because of bright moonlight, Kuiper mounted a binocular viewer on the instrument and trained it on the Moon. His observations led to several scientific papers, the first in 1954. The implication of Kuiper's conclusion was startling. It suggested that the Moon had been nearly completely melted by its own radioactivity some 0.5 to 1 billion years after its formation, and that the maria were formed during this epoch and not as the result of melting caused by the impacts. This conclusion that both impacts and the Moon's own internal heat were forces creating its surface features was novel at the time. Kuiper was not infallible, however, as his observations led him to believe, incorrectly, that crater central peaks were volcanic and that they formed only around the time the maria formed.

Kuiper's first ventures into the world of professional lunar research netted him a swat from Harold Urey, who took exception to Kuiper's statement that the Moon had melted. This set the stage for professional hostility between Urey, the cold Moon proponent, and Kuiper, the hot Moon advocate, which would last for years. Points and counterpoints aimed at each other's lunar views appeared in scientific journals.

Gerard Kuiper's interest in the Moon intensified in the mid-1950s, when he routinely observed the moon through the McDonald Observatory 82-inch telescope and recorded his observations on this German-language Moon map. This beginning eventually led to the establishment of the Lunar and Planetary Laboratory at the University of Arizona.

But just as Baldwin's lunar resolve was strengthened by his initial professional rejections, so did Urey's attacks strengthen Kuiper's resolve to do something significant with the Moon. The opportunity to do so came in the summer of 1955 at the Ninth Congress of the International Astronomical Union (IAU) held in Dublin. Kuiper's own observations of the Moon led him to the conclusion that, as dramatic as some were, existing photographs of the Moon were inadequate for proper lunar research. As president of IAU Commission 16, Physical Observations of the Planets and Satellites, Kuiper was positioned to do something about that. He called for ideas about how a new and more relevant lunar atlas should be assembled.

The sole response to Kuiper's atlas request came from Ewen Adair Whitaker, a self-taught English astronomer who had been interested in the Moon since 1951 through his work with the British Astronomical Association. Their meeting in Dublin led to Whitaker immigrating to America and working with Kuiper on the new atlas starting in October 1957. Kuiper had secured startup funding for his atlas from the National Science Foundation in April 1957. This was supplemented with contract money from the U.S. Air Force Cambridge Research Laboratory in the fall of 1957.

Lunar Studies Accelerate

Then, a new sense of urgency entered the centuries-long traditionally slow pace of lunar studies. While the world reacted in shock to the launch of Sputnik, the creation of a definitive lunar atlas now became a priority for Gerard Kuiper. At Ewen Whitaker's suggestion, another self-taught Briton was brought into the circle, David W. Arthur, who at the time was a photogrammetrist with the British Government's mapping agency. Arthur was also a member of the British Astronomical Association and so knowledgeable about the Moon that Kuiper asked him to write the selenography chapter for the fourth volume of *The Solar System*. In this, Arthur proved his worth as he apparently knew more about the Moon than his boss. In a sign of simpler times past, it did not bother the PhD. Kuiper that both Whitaker and Arthur lacked degrees. He was interested in their skills, not diplomas, and the team went on to produce many notable lunar research achievements together.

Kuiper's *Photographic Lunar Atlas* was published in April 1960 by the University of Chicago. The volume divided the Moon into 44 separate fields, which incidentally did not correspond with the 44 fields of the Lunar Astronautical Charts later prepared by the USGS. Each field in turn was covered by four to seven photographs taken under differing Sun elevations and differing librations. The images taken under differing librations could be viewed stereoscopically with a special viewer devised by Robert Hackman. All photographs were printed on the same 1:1,370,000 scale and presented the Moon's disk as 2.5 meters in diameter. The atlas, consisting of 230 prints, each 43 by 53 centimeters, were packaged in a big red box. It was the first widely available and scientifically useful collection of lunar photographs.

William Hartmann photographs a lunar image projected onto a white globe to achieve an overhead perspective along the lunar limb to create the innovative *Rectified Lunar Atlas*.

Because Kuiper's atlas was made possible in part by funding from the United States Air Force, it is sometimes called the *USAF Photographic Lunar Atlas*. In the late 1950s the Air Force assumed it would play a major role in the exploration of the Moon and funded academic research before Cold War politics pushed lunar efforts to NASA, a civilian space agency.

In December 1960, the first of four lunar atlas supplements was completed with the help of Arthur and Whitaker and released as the *Orthographic Lunar Atlas*. This large red book was created from selected images from the *Photographic Lunar Atlas*. Orthographic coordinates, or a grid that treats the Moon like a disk at an infinite distance, were added. Two versions of the atlas were released, one for the Air Force as the sponsoring body, and a commercial version. Edition A had an orthographic grid on each sheet while Edition B also had the selenographic longitudes and latitudes.

By 1960, Kuiper had initiated his Lunar and Planetary Laboratory at the University of Arizona in Tucson, where it remains to this day. At this new facility, the *Rectified Lunar Atlas* was compiled by Kuiper, Whitaker, William K. Hartmann, and L. H. Spradley. This large bound volume showed the lunar limb areas as if seen from overhead. The *Rectified Lunar Atlas* is famous for its role in the initial understanding of the impact ring structure of Mare Orientale.

The *Rectified Lunar Atlas* was not the first to apply photographic techniques to correct the foreshortening along the edge of the Moon. Between 1927 and 1940, Dr. F. E. Wright of the Geophysical Laboratory of the Carnegie Institution developed a technique of projecting lunar images onto a 45-centimeter glass

The *Rectified Lunar Atlas* used special photographic techniques to display the limb areas of the Moon as seen from overhead, revealing that the Cordillera and Rook Mountains on the western limb were the rim of a large circular impact basin cradling Mare Orientale.

globe that had been coated with photographic emulsion. When the globe was developed and illuminated from inside, the true shape of maria and other formations along the lunar limb was seen for the first time. At the University of Chicago, Harold Urey experimented with similar techniques in 1962 to secure images of Mare Australe that had not been covered in Wright's original work. Lunar images were projected onto a wooden globe that was coated with a white matte finish.

In 1967, Kuiper's team created the *Consolidated Lunar Atlas*, perhaps the best photographic lunar atlas published in the film photography era. It is considered both the third and fourth supplement in Kuiper's lunar atlas series as this work presented the Moon at both low and high solar elevations. Details as small as a kilometer are present in images taken with the 1.5-meter Catalina telescope near Tucson, with supporting images taken with a similar-sized instrument at the Naval Observatory's station in Arizona as well as the Yerkes and Lick refractors. The *Consolidated Lunar Atlas* quickly became a popular pre-Apollo lunar reference and copies were freely given out at lunar science conferences.

The Geology of the Moon

One of the last pre-Space Age big names to champion impact origin of lunar features was Eugene Shoemaker. In the early 1950s, Shoemaker realized rocket technology was advancing at a rate that would soon allow flights to the Moon. He also recognized our satellite as a world shaped by geologic processes and therefore a trained geologist should be one of the first scientists to go to the

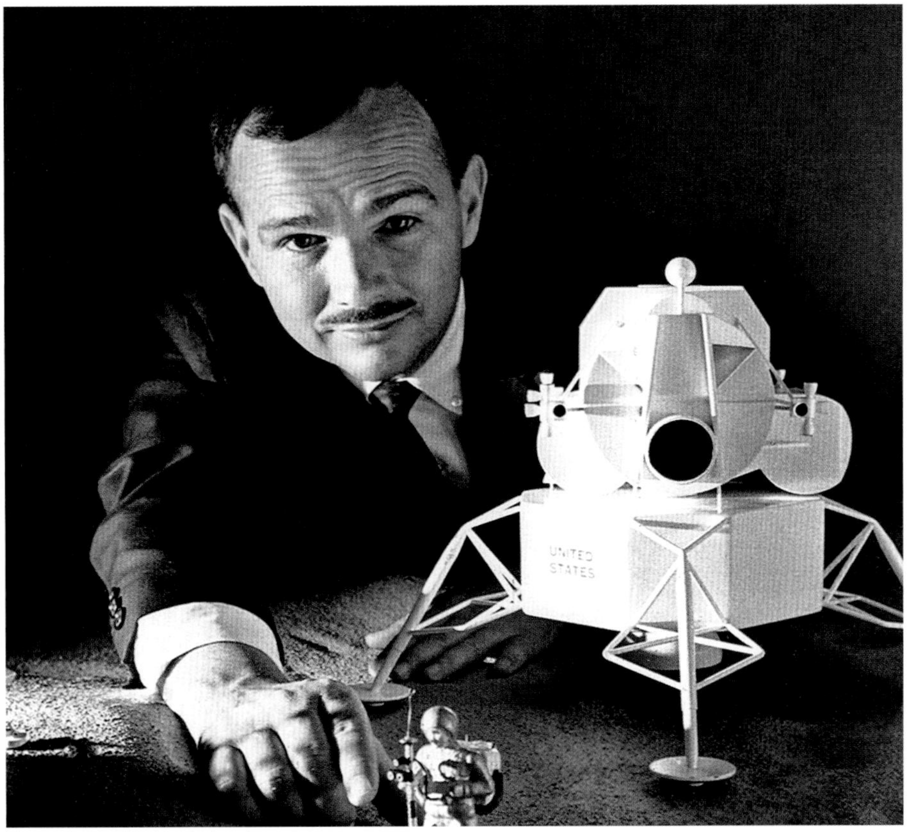

Eugene Shoemaker founded the astrogeology branch of the U.S. Geological Survey and aspired to be the first geologist on the Moon. Health issues prevented him from becoming an astronaut, but he played a key role in the geological training of the Apollo astronauts.

Moon. Shoemaker was determined to be that geologist and explore the strange world of the Moon.

To hone his astrogeology skills, Shoemaker carefully studied Meteor Crater and completed Gilbert's work of 60 years before. By comparing the Arizona crater with atomic bomb craters at the Nevada nuclear test site, Shoemaker proved beyond a doubt that Meteor Crater was an impact feature. Shoemaker also founded the astrogeology branch of the U.S. Geological Survey in 1961.

Another of Shoemaker's pioneering projects was mapping the sequence of the geologic ages of features near the Copernicus crater. Shoemaker's chart of the Copernicus area had its origins in a detailed image of the Copernicus crater taken by Francis Pease on September 15, 1919, with the 2.5-meter Mt. Wilson telescope. From this humble beginning, the Moon was systematically geologically charted over the next two decades by the USGS. The geologic mapping systems Shoemaker initiated carry over today into the charting of other planets.

Unfortunately, Shoemaker's space flight aspirations were cut short when he was diagnosed with Addison's Disease, an affliction easily treatable for daily life, but that medically disqualified him from astronaut training. Shoemaker then guided the USGS efforts in charting the Moon in support of Apollo and geologically training the Apollo astronauts who would go to the Moon. Although his own spaceflight plans were derailed, Shoemaker contributed more to the

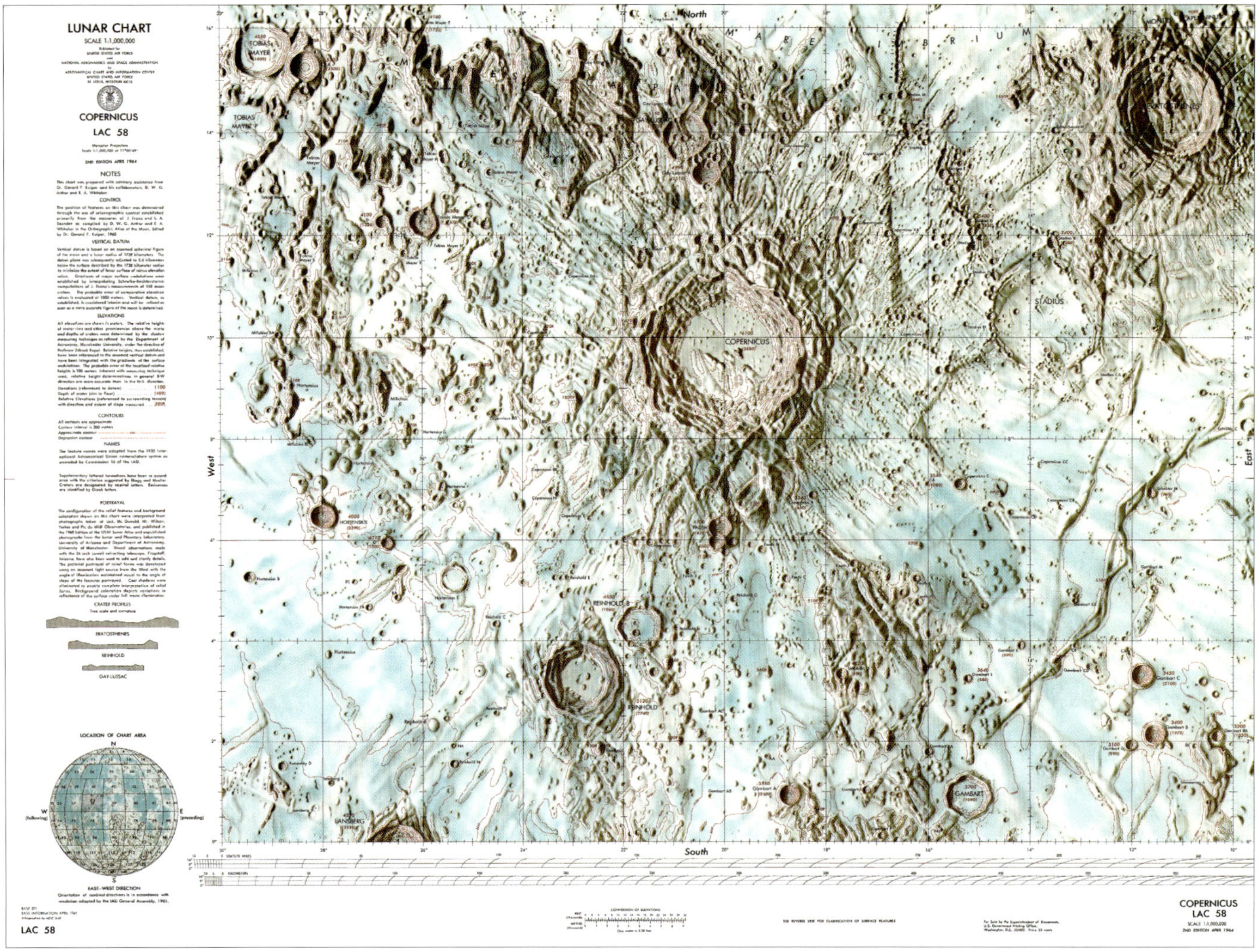

In preparation for upcoming lunar exploration, the U.S. Air Force mapped the face of the Moon in unprecedented detail, releasing a series of Lunar Astronautical Charts (LAC) that were the most detailed lunar reference before spacecraft photography in the late 1960s.

American lunar effort in a support role than he probably would have as an Apollo astronaut.

With the race to the Moon accelerating quickly, there was an urgent need for more detailed maps of the Moon. The Defense Department entered the lunar mapping field to provide the newly established NASA with the best lunar references possible. For the first time professional cartographers experienced at using aerial photographs were employed to chart the Moon.

In 1959 the U.S. Army Map Service began work on a two-sheet topographic map scaled at 1:5,000,000. The map was published in 1963 and compiled with 1,000-meter contours, with supplemental 500-meter contours derived by stereoscopic study of photos taken at extreme librations. At the same time, the U.S. Air Force Aeronautical Chart and Information Center also began charting the Moon at a scale of 1:1,000,000, resulting in a 3.5-meter lunar disk. Starting in 1961, visual observations with the Lowell Observatory 60-centimeter refractor were applied to the second edition of the atlas in 1964. By 1967, 44 charts measuring 56 by 74 centimeters, known as the Lunar Astronautical Charts (LAC) covering the central near side of the Moon were issued in collaboration with the University of Manchester, Pic du Midi Observatory, the Lunar and Planetary Laboratory, and Lowell Observatory. The LAC charts were soon supplemented by photography taken with the NASA 1.5-meter telescope at Catalina Observatory in Arizona and issued as the *Consolidated Lunar Atlas*.

The U.S. Air Force lunar mapping project was the last great Earthbound mapping effort before spacecraft took over the role. NASA's Lunar Orbiter missions mapped the Moon in exquisite detail between 1966 and 1968 in support of the Apollo program and rendered subsequent telescopic mapping efforts obsolete.

Other Earthbound lunar studies include a series of amazingly detailed images taken in June and July of 1962 through the three-meter Shane reflector at Lick Observatory. The breathtaking images were recorded on commercially available 10-by-12-centimeter Kodak Plus-X film plates and benefited from a period of unusually stable atmosphere at Mt. Hamilton. The images appeared in newspapers and magazines and several of the plates were published as two-page images in the popular *Sky & Telescope Magazine*.

Several notable atlases followed the series produced by Gerard Kuiper. An excellent but little-known atlas was produced at Kwasan Observatory by S. Miyamoto and his colleagues. A British effort in 1969 by H. R. Hatfield titled *The Amateur Astronomer's Photographic Lunar Atlas*, known as the "Hatfield Atlas," is not as detailed as professional atlases, but it was affordable by the amateur astronomer.

Accelerating interest in the Moon soon placed classic visual lunar observers on a collision course with the Space Age. Telescopic purists wanted to continue presenting the Moon with the south at the top of the map. The new breed of explorers planning to send rocket probes to the Moon wanted lunar charts to mimic terrestrial maps, with north at the top. A debate was carried out in the International Astronomical Union. Patrick Moore was among those calling for retention of the classic system, but the "north-toppers," as Moore called them, won the day. In August 1961 the IAU passed a resolution reversing the east–west directions on the Moon and north was placed at the top of all future lunar charts. Moore remained a holdout of the old system for years, thumbing his nose at north-toppers by stubbornly displaying lunar illustrations in his popular books with south up.

The Origins of Space Race Science

When the Apollo project became a national priority in the 1960s, the lunar exploration torch was passed from a few dedicated astronomers and geologists to engineers and industrial managers who guided an army of 400,000 workers in an effort to land a man on the Moon.

But signs of this onrushing epoch had been foretold for decades. Early in the 20th century, the Russian spaceflight theorist Konstantin Tsiolkovsky postulated the future of human habitation of the cosmos. In 1929 he wrote: "The conquest of the solar system will not only give us energy and life ...

Working in Cold War secrecy behind the USSR's Iron Curtain, Sergei Korolev was the anonymous mastermind of post-World War II Soviet rocketry and space exploration.

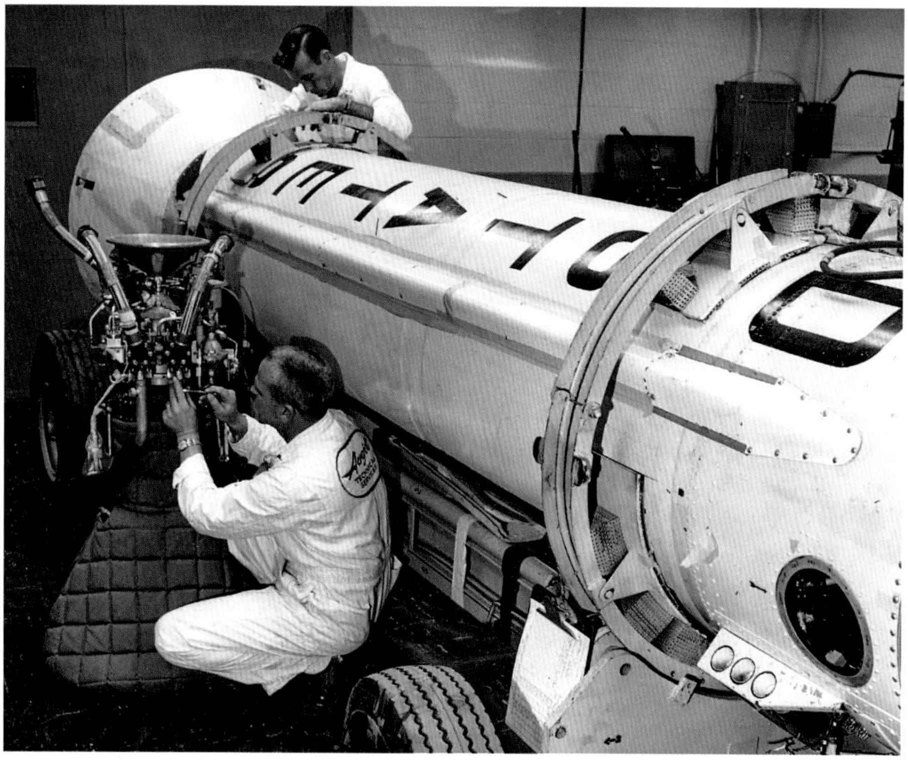

The initial Air Force Pioneer launches combined the second and third stages of the Navy's Vanguard launcher, shown here, with the Air Force's Thor ballistic missile, to create the Able booster.

in Los Angeles under the guidance of Brigadier General Bernard Schriever, and the Army Ballistic Missile Agency (ABMA) in Huntsville, Alabama under General John Medaris and with Wernher von Braun as technical director, would both mount missions to the Moon as quickly as possible.

ARPA's two-pronged assault on the Moon called for using proven existing military rockets. The Air Force was directed to prepare three Moon probes launched by the Able booster, a combination of an Air Force Thor missile and the upper stages of the Navy's Vanguard satellite launcher. The Army was to prepare two Moon probes using the Juno II which paired the Army's Jupiter missile with JPL's upper stages from the Juno 1. Pickering's Project RED SOCKS was becoming a reality.

Space Technology Laboratories, the contract manager for the Air Force ballistic missile program, and the Jet Propulsion Laboratory, long an Army contractor, were designated to work with the military in developing their respective lunar probes. Collectively, the ARPA Moon program was known as "Operation Mona." To the public, however, the American Moon shots were simply known as "Pioneer."

Many unknowns plagued the ambitious project. The three-stage version of the Able rocket had never been tried before. Achieving lunar flight called for maximum performance from the rocket and left no room for error. ARPA director Roy Johnson stated the project would achieve its goal if a probe approached within 80,000 kilometers of the Moon.

Spacecraft had never been tracked at lunar distances and hearing Pioneer's one-watt transmitter from the Moon would require extremely sensitive receivers.

A tracking network was quickly set up with sensitive antennas spread around the world. Existing antennas in Massachusetts, New Hampshire and Florida were supplemented by new tracking facilities built in Hawaii and Singapore. Even the newly commissioned giant 76-meter "Mk. 1" radio telescope at Jodrell Bank, in Manchester, England was brought into play to ensure 24-hour-a-day reception en route to the Moon. The network was completed just 17 days before the first Pioneer launch in the summer of 1958.

As the first Pioneer launch approached, public interest was high, but few details were released. Even the Jodrell Bank radio telescope, a British civilian installation, was tight-lipped about the Pioneer Moon plans. British newspapers openly discussed the presence of instrument trailers brought to the Mark 1 telescope that were stenciled "Jodrell Bank, U.S. Air Force, Project Able." However, Jodrell Bank director Bernard Lovell had to follow Air Force media guidelines that specifically stated, "Nothing was to be said by Jodrell Bank until the rocket had been launched." Even requests from other scientific institutions for the probe's radio frequency were replied with, "unable to answer without permission from U.S. authorities."

When the Soviets learned of the American Moon plans, their attention also turned in earnest to our nearest celestial neighbor. On January 28, 1958, rocket designer Korolev, along with Academician Mstislav Keldysh, sent a letter to the Central Committee of the Communist Party that outlined plans to send an initial probe to impact the Moon, then a later probe to fly around the Moon to image its far side. The plan was approved by First Party Secretary Khrushchev who, after the enormous propaganda success of the Sputniks, now took a political interest in space exploration.

Also spelled out in the letter to the Central Committee was an even more exotic plan originated by Academician Yakov Borisovich Zeldovich to explode an atomic bomb on the Moon. Zeldovich had been one of the original architects of the Russian atomic bomb along with Yulii Khariton and Igor Kurchatov. Zeldovich calculated that due to the lack of lunar atmosphere, there would be very little dust cloud visible after the initial flash, but the purpose of the lunar nuclear blast was primarily political. It was realized that after a brilliant atomic flash on the Moon, no one would question the reality of the Soviet Moon rocket. Anticipating the need for irrefutable proof that their craft had indeed reached the Moon, Soviet designers proceeded with the lunar atomic plan along with the non-explosive impact probe and far side photographic mission.

Work on the Soviet lunar atomic bomb progressed through building mock-ups. The device had the appearance of a floating naval mine with triggering rods protruding in all directions to initiate the explosion regardless of its impact angle. Fortunately, reason finally prevailed amid safety concerns. If the booster or core stage of the Moon rocket failed, the atomic device would fall back on their own land. If the third stage failed, the atomic explosives could fall anywhere on Earth with disastrous consequences. Without assurances of the rocket's reliability, Academician Zeldovich finally abandoned the project.

A Plan to Nuke the Moon

It bears noting at this point that serious planning to detonate an atomic bomb on the Moon was not solely a Soviet enterprise. Various informal studies prior to the Space Age mused about nuking the Moon for both scientific and political gain, but after the Space Race with the Soviet Union began in earnest in 1957, serious American military studies were initiated to formulate plans to detonate an atomic explosive on the Moon. Decades after the U.S./Soviet Moon race was over, these previously secret studies came to light.

The genesis of the lunar explosive idea came at the beginning of the 20th century from Robert Goddard, the father of modern American rocketry. In his famous paper *A Method of Reaching Extreme Altitudes*, Goddard discussed sending a rocket loaded with photographer's magnesium flash powder to the Moon. To establish the viability of his idea, Goddard even carried out a nighttime experiment near his Worcester, Massachusetts home in October 1916 to determine how much flash powder would be required to see a flash on the Moon. He ignited samples of magnesium in a glass vacuum bottle while observing the resulting flash from 3,600 meters away. Goddard determined the flash from 0.015 grams was readily visible. Goddard extrapolated his results to lunar distances and determined that if a modest 30-centimeter telescope was used to observe the lunar flash, 6.27 kilograms of magnesium powder would be easily seen.

In the 1940s, science writer Willy Ley expanded on Goddard's idea. Ley noted that if the lunar flash were obscured by bad weather, the Moon spectacular would pass unseen. He thus proposed the lunar impactor also carry half a kilogram of explosive to disperse 4.5 kilograms of white powdered glass that would leave a visible bright patch on the Moon.

In 1945 the meteoricist H. H. Nininger promoted the application of the newly developed guided missile and the atomic bomb to hurl lunar soil samples toward the Earth. His idea was an attempt to replicate the erroneous contemporary 1950s theory that the source of tektites, a type of melted glassy rock, was meteorite impacts or volcanic explosions on the Moon that hurled material to the Earth.

The RAND Corporation, a scientific think tank, began in 1956 to consider ideas about how to launch an atomic bomb to the Moon. The next year as space fever began to grip the United States, missile designer Krafft Ehricke and Nobel winner George Gamow proposed a small probe called Cow, so named after the nursery rhyme, to circle the Moon and return to Earth. A follow-on version was to be preceded by an atomic bomb that would throw up a cloud of lunar debris allowing the trailing probe to scoop up samples of Moon rock and return it to Earth.

In 1957, Gerard Kuiper, one of the leading astronomers of his era, coauthored a paper detailing the scientific aspects of detonating an atomic bomb on the Moon. Edward Teller, one of the scientists who created the hydrogen bomb, also speculated about launching atomic bombs to the Moon to produce lunar seismic data. The nuclear exploration of the Moon was further examined in proposals by the Jet Propulsion Laboratory in October 1957, as part of JPL's proposed Project RED SOCKS that envisioned nuclear excavation of surface material.

In 1958 the idea of sending nuclear explosives to the Moon gained military attention when the Air Force Special Weapons Center initiated project A119, "A Study of Lunar Research Flights." The project's goal was to impress the world with American technical knowhow with a nuclear flash visible by the naked eye from Earth. Engineers at the Lockheed Missiles and Space Division examined using the Agena upper stage developed for the Air Force Corona spy satellite program to propel an 11-kiloton nuclear bomb to the Moon. Among the scientists providing input to Project A119 was a young doctoral student named Carl Sagan, who would later gain international fame as an astronomy and humanity popularizer. Other scientists involved included Leonard Reiffel who would become the Apollo program's lunar mission scientific instrumentation manager.

The science goals of the nuclear project were outlined as studying the thermal characteristics of the lunar surface, examining any magnetic field and examining the seismic response to the explosion if several seismometers were first placed on the lunar surface. Carl Sagan's contributions were studies of how the nuclear cloud would respond to the Moon's lighter gravity and suggesting that the cloud be examined for organic material. Eventually, it was realized that the public opinion of a lunar atomic bomb would have been negative, and the project was terminated in January 1959.

Curiously, Carl Sagan's work on the A119 project could have gotten him in trouble in less scientifically frenzied times. He authored two papers studying aspects of lunar nuclear explosions: *Possible Contribution of Lunar Nuclear Weapons Detonations to the Solution of Some Problems in Planetary Astronomy* and *Radiological Contamination of the Moon by Nuclear Weapons Detonations*. Sagan cited these papers years later when applying for an academic fellowship. Since the papers were authored under the auspices of a then secret military project, some considered his mention of the papers as breach of national security.

As 1958 dawned, Korolev was now concentrating on Moon shots instead of Sputniks. Through his initiative, the first Soviet Luna probe, the Ye-1, was a lunar impact mission that was to explore near-Moon space en route to a free-fall crash on the lunar surface. The Ye-2 and Ye-3 photographic probes were to circle behind the Moon and relay photographs of the hidden side. The Ye-4 lunar atomic bomb project was still in the works but would soon be canceled.

The Race to the Moon Begins

Western observers had expected the Soviets to try for the Moon early in 1958. Instead, there was a half-year-long lull in Soviet space activity. The reason for the six months of apparent inactivity is now understood. The first two Sputniks were the result of the dedicated efforts of a very small band of technicians headed by Sergei Korolev. Unlike the infant American space program that in

In the aftermath of the first Pioneer's flaming end, American engineers asked the burning question; why haven't the Russians tried for the Moon? As rumors and speculation about Soviet Moon plans swept the world, Korolev and his team of scientists completed the development of the SL-3 booster and the first of the Ye-1 lunar payloads. As those preparing the Pioneers feared, Korolev's team made three Moon launch attempts from Russia's secret rocket base in Kazakhstan beginning in early fall 1958. All three launches were planned as lunar impact missions. After each failed, their existence remained a Soviet State secret.

On the afternoon of September 23, 1958, the first Soviet Moon probe mimicked America's Pioneer when the SL-3 Luna booster failed catastrophically 92 seconds after launch due to heavy vibrations in the booster.

Technicians at Cape Canaveral prepared Thor number 130 for the second American lunar attempt. This time, celestial alignments dictated a liftoff at 4:47 a.m., October 11, 1958. The new Pioneer's mission was identical to the ill-fated August effort, orbit the Moon. The rocket climbed perfectly into the starry Florida night. Hundreds of technicians and newsmen held their breath

In August 1958, the Able booster was erected on Cape Canaveral launch pad 17A in preparation for launch of America's first lunar probe.

Jet Propulsion Laboratory technicians, wearing surgical scrubs to prevent biological contamination, assemble the 51-centimeter-long Pioneer 3 probe.

as it passed the previous Pioneer's ill-fated 77-second mark. This time all was well. The booster continued to fire for its full 160 seconds, delivering the best performance of any Thor to date. The Able second stage then took over and thrust Pioneer further into the darkness. Finally, the third stage unleashed its brief but furious burst of energy and lofted the probe into deep space. Now officially christened Pioneer 1, the little spacecraft soared upwards to heights never before attained. The radio telescope at Jodrell Bank picked up signals from Pioneer 1 ten minutes after launch. For all appearances, this time America really was going to the Moon.

But, incredibly, Pioneer 1 failed to reach its goal. The plan had been to boost the probe to 10,744 meters per second. This would be enough to escape Earth's gravity, allowing Pioneer 1 to reach the Moon, where the probe could fire its retrorocket and enter lunar orbit. Several minor discrepancies together robbed Pioneer 1 of the precious final nudge of velocity: the aim was 2.1 degrees too high, a premature second stage shutdown, and the third stage was skewed slightly when it fired. These left Pioneer 1 approximately 251 meters per second short of the velocity needed to reach the Moon. The probe was doomed to loop 113,830 kilometers into space before dropping back to Earth. At 11:46 p.m., October 12, Pioneer 1 ended its life as a streak of flame above the South Pacific as it slammed into the atmosphere.

Even without reaching its celestial goal, Pioneer 1 did much to improve the sagging image of American space exploration. During its 43-hour flight into uncharted space, Pioneer 1 had achieved the highest velocity of any spacecraft and shot to an altitude 25 times greater than ever reached before. The probe also measured newly discovered radiation belts above 4,000 kilometers and made the first direct measurements of the Earth's magnetic field. Pioneer 1, a failed Moon probe, was thus known as the "magnificent failure."

Unknown to those furiously preparing Pioneer 1, another lunar countdown was also underway on the far side of the world. The next Soviet Moon launch occurred 15 hours after Pioneer 1 departed the Earth. As it turned out, the Soviet attempt fared worse than Pioneer. While the American probe looped one-third the way to the Moon, the Soviet booster exploded 104 seconds after liftoff. An investigation determined both Soviet lunar failures to date had occurred because adding the upper stage to the R-7 rocket raised the booster's center of gravity, thus setting up nine- to 13-hertz-resonance vibrations that tore apart

A dramatic time exposure records the service gantry rolling away from the Pioneer 3 prior to its launch on December 6, 1958.

the fuel lines feeding the rocket engines. The problem was solved by installing fuel line dampers to quiet the vibrations.

The third and last of the Air Force Thor-Able-powered Pioneers thundered up from Cape Canaveral at 2:30 a.m., November 8, 1958. After a flawless liftoff, Thor number 129 dazzled observers as it climbed above an unusual ground mist and barreled through a low-level cloud deck, illuminating the eerie scene with an orange glow.

After Thor's first stage performed perfectly, the second stage took over. Not wanting a repeat of Pioneer 1's premature second stage cutoff, ground control would signal the Able stage to shut down at the proper velocity. But the Moon eluded Pioneer. For some unknown reason, the third stage never ignited. After arcing upward to 1,550 kilometers, Pioneer 2 plummeted back to Earth above Northwest Africa.

The Army Aims at the Moon

After Pioneer 2's aspirations burned into cosmic dust, the lunar torch was passed from General Schriever's Air Force Ballistic Missile Division to the Army Ballistic Missile Agency. The next two Moon shots were now in the hands of the same team that launched America's first Explorer satellite. General John Medaris headed the Army team with the popular and charismatic German émigré Wernher von Braun as the chief technical advisor. William Pickering headed JPL's contribution to the Army Moon effort.

When Pioneer 1 discovered evidence of intense radiation at altitudes greater than the Explorer satellites could reach, scientific concerns again changed the payload of the little Army probes. The first Pioneer had only reached one-third the way to the Moon. Now, a crucial question needed to be answered. Was the radiation field local to the Earth or did it extend all the way to the Moon or even beyond? The very ability to send men to the Moon hung on this question. To find out, the lunar far side imaging experiment was dropped. In its place, two Geiger-Müller tubes were to register radiation levels between Earth and the Moon. One detector had a saturation point of 10 roentgens per hour, the other 100 roentgens per hour. The camera shutter trigger mechanism from the previous photographic payload was retained as an engineering test. When Pioneer passed within 32,000 kilometers of the Moon, the photocell would signal Earth.

The British 76-meter radio telescope at Jodrell Bank was also used for the Army lunar attempts. In an example of how the Pioneers were such a "hurry up and go" effort, JPL director William Pickering did not contact the bank requesting tracking help until just 17 days before the Army's planned launch date.

On December 6, 1958, all was ready for the Army's first lunar attempt. Unlike the first three Pioneers, the Juno II booster did not rest high on an impressive looking launch stand with a flame trench below. Pad 5 at Cape Canaveral was little more than a simple metal frame resting on flat concrete. As the Army rocket seemingly sat directly on the ground, here rested the lunar hopes of American scientists and a chance for American prestige to spring back from the embarrassment of losing to Sputnik a year before.

At 1:45 a.m., the countdown reached zero and the Juno II ignited. Smoke and flames burst out in all directions, obscuring the stubby rocket as it slowly rose. Pioneer 3 seemed to be on its way toward the Moon. But again, this was not to be. The Juno was supposed to climb at an inclination of 68 degrees while its main engine burned for 179.8 seconds. Soon it became apparent the booster was climbing at between 71 and 71.5 degrees inclination. At this rate, Pioneer 3 would miss the Moon by too wide a margin for the photo-trigger to work. Then came Pioneer 3's death knell, the main engine exhausted its fuel 3.7 seconds too soon and flamed out. Although the upper stages performed flawlessly, they could not make up a 316 meters per second velocity shortfall. The Juno rocket's push was sufficient to zoom Pioneer 3 some 107,264 kilometers into space before it lost its battle with Earth's gravity. The Army's first Moon probe lived for 38 hours, 6 minutes, then flamed like a meteor as it burned in the atmosphere over French Equatorial Africa on December 7.

Although not a success, Pioneer 3 vindicated the decision to remove the photographic equipment and substitute radiation detectors. The probe confirmed a radiation belt around the Earth at 4,800 kilometers altitude and discovered a second radiation belt 11,250 kilometers beyond the first.

After NASA was created in mid-1958, it inherited the Pioneer program from ARPA. In its infancy, the new civilian space agency was little more than a spectator watching the program continue under its own inertia. When Pioneer 3's booster lifted off from Cape Canaveral with the NASA logo emblazoned on its side, the agency was still just a national figurehead for space exploration.

The Soviet Ye-1

Two days before the American Pioneer 3 left for the Moon, the third 1958 Soviet lunar launch took place. The December 4 launch performed normally until the SL-3 core stage engine lost thrust after 245 seconds of flight. This time the problem was the failure of the 5,100-horsepower hydrogen-peroxide-powered turbine that drove the fuel pumps on the second stage RD-108 engine. As usual, the Soviet failure remained a state secret while the hapless Pioneer 3 lived its brief life in full view of world publicity.

The day after New Year 1959, the USSR hurled a scientific bombshell. Radio Moscow proclaimed the beginning of a new age for human exploration: "On January 2, 1959, a cosmic rocket was launched toward the Moon. The launching again demonstrates to the world the outstanding achievements of Soviet science and technology." The startling announcement heralded Luna 1, the world's first spacecraft to reach what the Soviets called the "second cosmic speed" or escape velocity. On their fourth lunar attempt, a Soviet Ye-1 probe had slipped the bonds of Earth's gravity and left our world forever.

The 361-kilogram Luna 1, called "Mechta" (The Dream) by the Russians, was launched on January 2, 1959. When the third stage burned out 12 minutes after liftoff, the vehicle was 1,500 kilometers above Siberia and climbing

Luna 1, the first spacecraft to escape Earth's gravity, lifted off from the Baikonur Cosmodrome in Kazakhstan on January 2, 1959. Intended as a lunar impact mission, a guidance error resulted in a lunar miss and the spacecraft, called Mechta, entered solar orbit as an artificial asteroid.

Mechta missed the Moon, but was nonetheless a propaganda victory for the USSR when the 361-kilogram probe escaped Earth's gravity.

toward the Moon. Thirty-four hours later, at 5:59 a.m. Moscow time, January 4, Mechta passed within 5,955 kilometers of the Moon. Streaking onward at 2.45 kilometers per second relative to Earth, Mechta remained in contact for 62 hours until its batteries depleted on January 5 while 500,000 kilometers from Earth. Mechta's planned mission was to impact the Moon. The probe missed its target and entered a 446-day solar orbit ranging from 146.4 million kilometers to 197.2 million kilometers from the Sun. On this path, the probe approaches the Earth every 5.43 years.

According to Alexandre Ananov, the Russian founder of the International Astronautical Federation, Mechta's booster imparted an excess of velocity, causing the lunar miss. Soviet technicians later revealed Mechta had missed the Moon because of "human error in calibrating a direction finder." What the cryptic explanation meant was the lunar miss was created by a two-degree error in setting the radio guidance system used with the R-7 missile that formed the first and second stages of the Luna booster.

The electronics in the historic Ye-1 probe were housed in an 80-centimeter diameter spherical shell made from polished aluminum-magnesium alloy and pressurized to 1.3 atmospheres. A one-meter magnetometer boom and four symmetrically placed rod antennas protruded from one hemisphere. Mechta also carried two piezoelectric micrometeorite detectors, cosmic ray and plasma detectors and internal pressure and temperature measuring instruments. Both Mechta and its booster carried a commemorative 9-by-15 centimeter "imperishable plaque" reading "USSR, January 1959" in Cyrillic script.

The Luna probe was ejected from its launcher so the booster's metal body would not interfere with the instruments. Scientific data was transmitted in 50.9-second bursts at 19.993 MHz while the probe received signals from Earth at 183.6 MHz. Additionally, the final stage booster rocket transmitted cosmic ray data in 81.6-second bursts at 19.995 MHz and 19.997 MHz. Listeners at the Moscow Planetarium were able to tune in Mechta by using short-wave radio equipment.

A significant scientific result of Mechta's flight was the detection of a zone of intensive radiation between an altitude of 5 and 6 Earth radii. This supplemented previous measurements of the outer Van Allen radiation belt that surrounds the Earth.

Soviet deep space tracking capabilities were crude in 1959. To help pinpoint the location of the cosmic rocket as well as study the properties of translunar space, a spectacular sodium "artificial comet" experiment was prepared by well-known Soviet astronomer Professor Iosif Shklovsky. A quartz clock timer ignited one kilogram of sodium carried by Mechta's booster at 3:57 a.m. on January 3 while approximately 120,000 kilometers from Earth. The sodium evaporated in six seconds and fluoresced in sunlight for five minutes, creating a 650-kilometer diameter artificial comet that was visible in the constellation Virgo from Earth's Eastern Hemisphere. The spectacle was a propaganda coup, leaving no doubt that Mechta was real.

Still, many American officials remained skeptical about the probe. Some openly speculated about Soviet fraud. Long-range radars in Turkey that tracked suborbital shots from Russia's Kapustin Yar missile base did not detect Mechta's liftoff, because it launched from the secret rocket base known today as the Baikonur Cosmodrome. Only a single U.S. tracking station in Hawaii received any indication that a Soviet launch had taken place on January 2, but at that time the nature of the launch was unknown. The Pentagon learned of the Soviet Moon launch only when it came across wire service news tickers.

In a patriotic rush, Soviet scientists hailed Mechta as the "10th planet," but it was only legitimately considered an artificial asteroid. The press nicknamed the probe "Lunik," and the popular name caught on and for several years afterward, the word Lunik was synonymous with any Moon probe.

Maximizing every propaganda opportunity, Radio Moscow made no mention that the craft missed its intended lunar target. Soviet propaganda rhetoric hammered the theme that Mechta was the result of "the creative toil of the whole Soviet people (in) the development of Socialist society and in the interests of all progressive mankind." Premier Nikita Khrushchev touted the space spectacular by noting Russian scientists "are the first in the world to map out the way from the Earth to the Moon." Khrushchev also said he felt "like hugging the man who has produced this, the first cosmic rocket, a new victory for the Soviet Union." Whether Korolev ever received his hug is not noted in history. Khrushchev's love of propaganda bluster did, however, forever lock his image as part of the international effort to reach the Moon early in the Space Age.

At a time when Cold War tensions were high, the Soviet Moon flight was a demoralizing blow to the American space effort which had tried to reach the Moon four times and failed. Mechta further needled American scientists with a salvo of impressive statistics. Pioneers 1 and 3 each reached only a third of the way to the Moon before falling back to Earth. Their masses paled in comparison to Mechta, which was heavier than all combined American space probes to date. A final insult was the Soviet rocket's trajectory was three times more accurate than the Pioneers.

As Mechta receded from Earth, President Eisenhower directed a congratulatory statement toward those responsible for the mission. But, despite Eisenhower's public praise for the Soviet Moon shot, Mechta left Washington officials in a less than cheery mood. A hot topic of debate in Washington was whether the "Lunik" was a hoax. Lack of definitive tracking left many skeptical and the American public had a hard time believing Mechta had passed the Moon. At the time of Mechta's flight, just 15 months after the first Sputnik, there was no experience tracking spacecraft beyond low Earth orbit. The Jodrell Bank Observatory failed to pick up any signals from the craft and reception by other stations was hit or miss. One problem was the USSR's choice of radio frequencies. Mechta transmitted on several bands. One at 183.6 MHz was very close to the commercial television frequency spectrum and existing space tracking stations were not configured for this frequency. Mechta's other frequency was near 20 MHz, a band sensitive to interference from terrestrial stations and ionospheric noise. The ionosphere is "quieter" at night, but during Mechta's

The Jet Propulsion Laboratory's network of ultrasensitive 85-foot steerable dish antennas were utilized to receive Pioneer's 0.1-Watt radio signal at lunar distances.

flight, the craft was in the same region of the sky as the Moon. This limited good reception on this band to five hours per day between moonrise and sunrise.

By the time Luna 1 came into view of U.S. stations it was already 170,000 kilometers from Earth. The only non governmental reception was by Stanford University in California, which received signals for two hours and 15 minutes on January 3, 1959. By working night and day, technicians at the Jet Propulsion Laboratory tracking station in Goldstone, California, were able to reconfigure the receiver on their 26-meter dish antenna to receive Mechta's signals. By the time JPL heard the craft at 183.4 MHz on January 4 it had already passed the Moon. An American station in Hilo, Hawaii, also heard Mechta and evidence mounted to counter claims by naysayers who asserted that Mechta was a hoax.

The flight of Mechta caught the world's imagination and provided the political impetus to direct the U.S. space program into an almost Moon-only objective in a race with the Soviets. Politics and international prestige were now more important objectives than the acquisition of scientific data.

Pioneer 4

Russian space prestige reached an all-time high with the flight of Mechta, but Operation Mona had one chance left to recoup American scientific honor. At 1:45 a.m., March 3, 1959, the Army's Juno II again sprang to life and soared into space. This time, less than four minutes after liftoff, Pioneer 4 was really on its way toward the Moon. The spacecraft was to become the second artificial asteroid, but the occasion was anticlimactic as it lay in the shadow of Mechta. For that reason, Pioneer 4 was tagged the "Me too" probe as it followed the Soviet spacecraft around the Sun.

Pioneer 4 had escaped Earth forever, but its path was not the one chosen by scientists. When Pioneer 4 arced upward from Cape Canaveral, Jodrell Bank picked up its signals 13 minutes after launch. After two hours of tracking, it became apparent the probe was not on the desired course to pass near

Technicians make final adjustments to Pioneer 4 atop its fourth stage rocket before the Juno II nose cone is installed.

the Moon. Director Lovell announced to the press that Pioneer 4 was "off course." The Juno II had boosted the tiny probe to 317 kilometers per hour faster than escape velocity, but it was still 303 kilometers per hour shy of the desired velocity. This, coupled with a 4.5 degree elevation error and 1.3 degrees azimuth error by the Juno II, meant Pioneer 4 would miss the Moon by 60,000 kilometers.

Nearly two days after launch, Pioneer 4 crossed the Moon's orbit while traveling 7,225 kilometers per hour relative to Earth. On its way to the solar system, the probe relayed important radiation data. It found that the lower Van Allen belt consisted of high-energy protons and low energy electrons while the upper belt consisted of low energy protons. More important, Pioneer 4 showed that beyond the Van Allen belt, radiation levels were low enough for humans to safely travel to the Moon. Eighty-two hours after launch, Pioneer 4's last whisper was heard by the Goldstone tracking station as the probe receded beyond 663,000 kilometers from Earth. The silent probe then slipped into a 397.75-day solar orbit ranging from 147.6 million kilometers to 173.7 million kilometers.

Von Braun and Pickering basked in the public attention given to the U.S. space program following Pioneer 4's flight. But the success was bittersweet. Mechta had been first to get near the Moon and well into the second year of the Space Age, America was perceived as playing "catch-up" with Russian space efforts.

Ironically, Operation Mona was a total lunar failure. Not one scrap of new information about the Moon had been gained. However, in the fledgling days of the Space Age, the very act of trying to get to the Moon was heroic and a source of invaluable engineering experience. The high flight of Pioneer 1 and the solar orbiting Pioneer 4 were also vital to U.S. prestige and showed America could respond to the scientific and propaganda challenges of its Cold War adversary. Operation Mona's race for the Moon with the Soviet Union had set into motion a series of lunar explorations that would continue until astronauts walked on the Moon 10 years later. In this respect, Operation Mona was an unqualified success.

The Soviet Luna 2 Reaches the Moon

As the first American lunar efforts concluded in spring 1959, the next phase of U.S. lunar exploration began. Substituting the larger Atlas missile for the Thor, the Air Force prepared the Atlas-Able booster to send the next generation of heavier Pioneers to the Moon in late 1959. Because Mechta had missed its target, the USSR had not actually reached the Moon yet. America hoped it still had a chance to claim the Moon prize.

The Soviets tried again for a lunar impact mission on June 14, 1959, but oddly, this lunar launch attempt was delayed for two days when the second stage was accidentally filled with the wrong fuel. When launched on June 16, 1959, the SL-3 core stage gyroscopic guidance failed 153 seconds after launch and the booster was destroyed.

Then came the galvanizing event the Western world feared would come. On the evening of September 12, Radio Moscow interrupted a program of popular music and announced a TASS news agency bulletin: "A second successful launching of a cosmic rocket took place in the Soviet Union today. The last stage of the rocket, having exceeded the second cosmic speed of 11,200 meters per second, is flying toward the Moon."

The 390-kilogram Ye-1 spacecraft, now called Luna 2, was launched from Baikonur into a direct-ascent trajectory toward the Moon. Before the rocket's precise trajectory was known, the Soviets hedged their bets when announcing the flight. In the Russian language, the preposition "k," pronounced "kuh," means both to and toward. Thus, the launch announcement was not clear about the mission's planned outcome. Later, while Luna 2 was still in flight, the Soviets refined the announcement to say that it would impact the Moon near the junction of the Sea of Serenity, Sea of Tranquility and Sea of Vapors. The probe, the first object to strike another celestial body, was thus aimed square between the Man-in-the-Moon's eyes.

Luna 2's instruments were basically identical to those carried by Mechta the previous January. However, the new probe's radio systems were improved to prevent the weak signal and poor reception that plagued Mechta's communications. The probe transmitted on frequencies of 19.993, 19.997, 20.003, 39.986 and 183.6 MHz. Luna was heard by many amateur radio operators in Europe and Asia, and this time, American tracking stations had no problem tuning in the Moon-bound craft. The Soviet Union still had no worldwide tracking facilities and officials requested that tape recordings of Luna's signals be mailed to the address of "Moscow, Cosmos."

Academician Leonid Sedov said the new Luna also carried several versions of a commemorative "imperishable plaque" based on the ideas of Tsiolkovsky. Within Luna was a 90-millimeter diameter ball designed by Arvid Pallo. The sphere was made up of 72 pentagonal sections, each of which displayed the

Luna 2 and its third stage rocket carried 9- and 15-centimeter diameter metal spheres that were designed to shatter on impact and scatter commemorative medallions across the impact site.

radiation particles than there are at 400 kilometers. Above 60,000 kilometers, Luna reported that cosmic radiation was "negligible," only two particles per second per cubic centimeter, and did not increase near the Moon.

The plasma ion traps carried by Luna 1 and 2 were designed by Konstantin Gringauz. The instruments were hemispherical in shape; thus they measured the flow of plasma in only one direction. The signal from Luna 1 displayed a periodic fluctuation as the probe slowly tumbled through space. Before publishing the results, Gringauz changed the ion traps installed on Luna 2. Instead of coplanar, the four traps were now in a tetrahedral arrangement. Once Luna 2 was outside Earth's magnetic field, it confirmed the directional flow of plasma in space. Thus, on Russia's first successful deep space probes, instruments designed by Konstantin Gringauz discovered the solar wind.

Luna's micrometeoroid detector recorded 0.002 impacts per square meter of area by high velocity particles weighing less than a millionth of a gram. While Luna traveled to the Moon, it detected less than 100 gas molecules per cubic centimeter of space. Approaching the Moon, it registered an increase in ionized gases at a 10,000-kilometer altitude. This led the Soviets to wonder if the Moon had a tenuous ionosphere.

The magnetometers carried by both Luna 1 and 2 were designed by Shmaia Shlemovich Dolginov. The devices detected no lunar magnetic field down to an altitude of 55 kilometers and found no lunar radiation belt. Dolginov is thus credited with discovering the Moon has no significant magnetic field. The lack of a magnetic field led American lunar expert Gerard Kuiper to reason that the Moon's core must be solid and not molten like Earth's. Harold Urey, however, speculated that the Moon may well have a liquid molten core, but its slow rotation rate does not move the core fast enough for a dynamo effect to create magnetism.

Many people, including some government leaders in the United States, refused to believe that Luna 2 had impacted the Moon. Some suggested that Luna 2 carried a clock timer to shut off the signal, simulating a lunar hit. Tracking information gathered by the Jodrell Bank radio telescope in England refuted this naive view. The 76-meter diameter instrument positively established Luna 2's velocity increase near the Moon by Doppler tracking and confirmed the lunar impact. When the signals ceased, the radio telescope was pointed directly at the Moon.

Two days after Luna reached the Moon, Jodrell Bank sent a message countering claims that the Moon strike was a hoax. It read: "Position of rocket signals did not agree with that of the Moon until the end of period. Doppler record proves the rocket did hit the Moon. J. G. Davies, Jodrell Bank."

The launch of Luna 2 came just three days before a state visit to the United States by Soviet Premier Khrushchev. When he met with President Eisenhower, Khrushchev gleefully presented him with replicas of the commemorative medallions carried by Luna 2. Khrushchev told Eisenhower: "We have no doubt that the excellent scientists, engineers and workers of the U.S.A. who are engaged in the field of conquering the cosmos will also carry your pennant over to the Moon. The Soviet pennant, as an old resident, will then welcome your pennant,

and they will live there together in peace and friendship." Khrushchev's eloquent words were lost in the media by the publicity from the fit the Russian statesman later threw over being denied a visit to Disneyland in California. Khrushchev's desire for a little capitalistic fun was derailed by concerns over his security.

When Luna 2 struck the lunar surface, it concluded the first chapter in man's rocket exploration of the Moon. Pushed by Cold War politics and national pride, a messenger from humanity reached our natural satellite a mere 15 months after the Space Age began. Russia had won a lunar battle, but the war was far from over. No one in their wildest dreams imagined that because of the race between the Pioneers and the Lunas, humans would voyage to the Moon just 10 short years later.

The launch of Luna 2 coincided with a State visit to the USA by Soviet Premier Nikita Khrushchev, where he presented a replica of the Luna 2 medallion sphere to Vice President Nixon and President Eisenhower.

3 The Moon Is a Harsh Mistress

While the American lunar effort continued in late 1958, preparations began for the initial planetary explorations. By mating the two upper stages from the Thor-Able Moon rocket to the more powerful Atlas booster, the newly created National Aeronautics and Space Administration (NASA) could loft a 122-kilogram Pioneer probe toward the planet Venus in June 1959. But the success of Mechta created such a political sensation that the Venus probe was dropped, and all the Atlas-Able probes were retargeted toward lunar orbit.

The new spin-stabilized Pioneers were giants compared to the previous tiny American lunar craft that could be cradled in one's arms. The one-meter-diameter spherical aluminum craft had grown to 169 kilograms. Four solar panels, each a half meter square, extending from the spacecraft's equator gave the craft the nickname of the "paddle-wheel satellite."

Technicians assemble the 169-kilogram Pioneer P-3, the first of NASA's ambitious, but ill-fated, lunar probes launched by the Atlas-Able rocket.

A static test firing of Atlas 9C slated to send Pioneer P-3 to the Moon went terribly wrong on September 24, 1959. A premature engine shutdown ruptured the fuel lines, sparking a fire that engulfed the vehicle resulting in an explosion 20 seconds later.

Scientific instruments aboard included a television scanner to image the far side of the Moon, radiation detectors, a magnetometer and micrometeorite detectors. Data was to be sent to Earth over twin 1.5-watt transmitters operating at 378 MHz. Protruding from the spacecraft's poles were twin hydrazine-fueled thrusters. The one at the base could fire four times to increase the probe's velocity if needed, while the one on top could fire twice to slow it down to enter lunar orbit. The new Pioneers were the first lunar probes with a self-contained liquid-fueled propulsion system.

Just 10 days after Luna 2 struck the Moon, the Atlas-Able booster underwent a crucial pre-launch test where the vehicle's engines were static-fired on the launch pad to validate their reliability. The September 24, 1959, test had disastrous results. The Atlas exploded and destroyed the probe, ending any hope of quickly following Luna 2 with a more advanced American spacecraft.

Then, less than a month after striking the Moon, the USSR made another startling announcement. At 1:00 p.m., October 4, 1959, Radio Moscow broadcast: "Attention, attention, dear Comrades! Listen now to the signals coming from the cosmos, from the third cosmic rocket launched today." What followed sounded like a raspy violin playing the A note above middle C. The source of the signals, Luna 3, had been launched from Baikonur by an SL-3 booster on the second anniversary of Sputnik.

The Soviets again provided Jodrell Bank with frequency and tracking data for the Luna 3 flight. The big Mark 1 antenna heard signals for 20 minutes, about 10 hours after launch, then they abruptly shut off. The Soviet Union announced the new Moon probe would transmit two to four hours per day on command from Earth while in view of Soviet territory. Jodrell Bank subsequently received data from the craft several hours each day on October 4, 5, 6, 7 and 9.

Early announcements were vague about Luna 3's objective, but unofficial news reports were specific. Georgy Duboshin, head of the celestial mechanics

A replica of the Luna 3 photographic probe at the Memorial Museum of Cosmonautics in Moscow details the complexity of the first spacecraft to explore the far side of the Moon in October, 1959.

department of Moscow's Sternberg Astronomical Institute stated the probe would photograph the far side of the Moon. It was not until October 19, over two weeks later, when the Soviets had the pictures in hand, that official sources admitted that Luna had indeed accomplished this.

Luna 3, officially designated as a Ye-2A spacecraft, was designed by Gleb Maksimov and carried a radio system designed by Evgeny Boguslavsky. The 278.5-kilogram photographic probe was a short cylinder 120 centimeters in diameter and 130 centimeters long with dome-shaped ends. The craft was pressurized to 1.35 atmospheres and had portholes on both ends for the cameras and Sun and Moon seekers to view their targets. Luna 3 was the first Soviet spacecraft powered by solar cells and was the first to be three-axis-stabilized using a system developed by Boris V. Rauschenbach that used gyroscopes and gas jets located on the end opposite the camera.

Luna 3's Yenisey-2 camera system used a 200-millimeter focal length f/5.6 wide field lens and a 500-millimeter focal length f/9.5 narrow field lens. The optical components for the Yenisey-2 camera system were manufactured at the Krasnogorsk Mechanical Plant, a facility that manufactured, among other things, aerial reconnaissance cameras. To ensure details were recorded over a variety of lunar albedo, the camera shutter was cycled between 1/200th, 1/400th, 1/600th, and 1/800th of a second. Small-scale images were 10 millimeters in size on the film and showed the entire Moon while large scale were 25 millimeters.

Soviet Moon Photos Taken with American Film

In an odd twist, the film chosen to photograph the far side was not Soviet; it had been salvaged from an American Genetrix spy balloon that had crashed in Soviet territory. Starting in 1956, prior to overflights by the CIA's U-2 spy plane, America's spymasters floated 516 high altitude camera-carrying balloons over Soviet territory. Floating at altitudes too high for interception by jet fighters, the jet-stream-borne balloons would take as long as a week to cross the Soviet Union automatically snapping pictures on the way from Europe to the Pacific Ocean. Once beyond USSR territory, American spooks would recover the film in midair with specially equipped aircraft.

Soviet film manufacturers were unable to create film sufficiently resistant to cosmic rays to ensure the Moon-bound film would not be spoiled. Thus, the American spy film, designed to function in the low density and temperature extremes of near space aboard the high-altitude balloons, was chosen instead. The salvaged Genetrix film was recut and perforated to the 35mm format for use in the Yenisey-2 apparatus. Though politically stung at the time by the Soviet Moon triumph, Kodak engineers who worked at the birth of the Space Age could later silently smile with pride knowing the film that took the first images of the far side of the Moon came from Rochester, New York, and not Moscow. The film was automatically developed aboard the Ye-2A spacecraft shortly after exposure by an apparatus designed to work in a weightless environment. A facsimile system then scanned the film at various resolutions up to 1,000 lines per frame. The photographs were transmitted at 39.986 MHz and 183.6 MHz through redundant radio systems.

Official details of the Luna 3 mission as described by Soviet officials paint a picture of a well-planned scientific program to explore the far side of the Moon. However, the true conduct of the Luna 3 explorations bordered on being haphazard and ill-planned. The hidden details of the Luna 3 operations were uncovered decades after the historic mission by space historian Don P. Mitchell through extensive translations of Soviet technical journals and the cultivation of friendships with the principals involved in the early Soviet space program.

Compared to NASA mission tracking standards, and the scientific importance of the far side photographic mission, the receiving station cobbled together for the Luna 3 flight almost seemed primitive. Located on the slopes of Mount Kosh-Kaya near Simeiz in the Crimea facing the Black Sea, the station was described as set up on a "steep seashore." The station was prepared by Yevgeni Boguslavsky in cooperation with military unit 32103 and became operational on September 23, 1958, the launch date of the first Ye-1 lunar impact mission. In preparation for the more advanced Ye-2 photo mission, the receivers and photo recording apparatus were transported by aircraft to the Crimean tracking site and a backup location in Yelizovo on the Kamchatka Peninsula. Arriving on October 4, Luna 3's launch day, the equipment was set up and tested over the next several days.

The antenna for communicating with Luna 3 consisted of a 100 square meter array of crossed dipoles that resembled a huge gridwork. The receiving antennas consisted of two stacked three-element Yagi antennas for reception

Soviet photographic engineers were unable to create radiation resistant film for Luna 3's cameras, so Eastman Kodak film salvaged from an American Genetrix spy balloon that crashed in USSR territory was used for the Luna 3 lunar far-side photography.

at 19.993 MHz and a crossed Yagi for 39.986 MHz. These antennas were each mounted on a simple wooden pole at the shoreline of the Black Sea. In retrospect, amateur radio operators often have more complex antennas in their backyards. The theory was the antennas would receive both the direct signal from the spacecraft and the signal reflected off the water. The receiving equipment was housed in trailers and the Luna 3 command center was set up in a temporary wooden building. One room housed the recording equipment including a tape recorder, photographic recorder, and a thermographic paper plotter for printing real time images. When in operation, the plotter was described by a mission participant as having a flying purple spot that scanned across the paper, slowly building up an image line by line. The station staff lived in tents and were fed by a military field kitchen. Compared to NASA's worldwide network of massive steerable dish antennas and modern electronic control rooms, the Luna 3 receiving station was incredibly primitive. Additional details about the reception of the Luna 3 far side images were revealed by Swedish space historian Sven Grahn.

Preparations for Luna 3 were hurried right up to launch. Only at the launch site was the craft's electrical power system finally adjusted properly. However, soon after launch, radio signal reception from the Moon probe was very poor because of design deficiencies and the primitive state of Soviet electronic communications. The Luna 3 radio system used a carrier wave system instead of a more efficient pulsed system. Luna 3's electronics were transistorized, which by contemporary Soviet standards, was a risky design strategy. The system's designer, Boguslavsky, blamed the poor reception on the craft's low transmitter power.

On the morning of October 5, not even the huge 76-meter Mark-1 dish in England could reliably receive signals from Luna 3. Jodrell Bank noted the frequency quickly fluctuated and at 13:51 Universal Time the frequency started drifting downward rapidly. The British receiver was 10 times as sensitive as the Soviet receivers and the mission was clearly in trouble.

The Desperate Salvage of Luna 3

The degree of Chief Designer Sergei Korolev's micromanagement is demonstrated by his decision that, instead of letting his engineers work on the radio reception problem, Korolev and other senior managers, including spacecraft systems designer Arkady Ostashev and Academician Mstislav Keldysh himself, should immediately travel to the tracking station in the Crimea and personally handle the poor radio reception problem. Korolev called his deputy, Boris Chertok, and told him, "Come along, bring no documents, don't count on returning home today." The race to the Crimea to save the mission was on.

Korolev and his team had to be at the Crimean tracking station by 4 p.m. when Luna 3 would come into radio range. The group departed Moscow at 10 a.m. in a hurried car trip to Vnukovo Airport and a flight to Simferopol. There, they boarded a helicopter for a flight to Mount Kosh-Kaya, but a sleet storm forced them to divert to Yalta, where cars were waiting to complete their trip. In what has been described as a hair-raising high-speed dash in a Zim limousine, the team arrived in time to begin analysis of why Luna 3 was misbehaving.

Korolev quickly determined the problem was two-fold. Luna 3 was overheating in the constant direct sunlight and flight controllers were not yet skilled in remotely operating machines in deep space. Korolev found an intermediary layer of bureaucracy in the form of Army officers between the spacecraft controllers and the spacecraft communicators. The root of Luna 3's control problems was lack of verification that certain operations had been completed before initiating new ones. Fortunately, Ostashev was able to cool the spacecraft by orienting it lengthwise to the Sun and programming various heat-producing devices to turn off when they were not needed, allowing Luna 3 to continue its mission.

Amazingly, the designer of the Yenisey-2 imaging system remarked the day before the lunar photos were to be received that there might not be sufficient film in the photorecording apparatus to complete the mission. Korolev was furious that he was not informed of this so he could have brought film with him when he flew in from Moscow. A separate jet was dispatched from Moscow to Simferopol with more film, which was then brought to Mount Kosh-Kaya by helicopter.

On October 6, 1959, Luna 3 passed 7,000 kilometers south of the waxing crescent Moon allowing photography of the eastern portion of the Moon's far side. For tracking stations in the Soviet Union to receive signals from Luna 3, the craft had to return to Earth over the north pole. A traditional launch to the Moon from northern latitudes would return the probe over Earth's south pole. The solution was to use a gravity assist trajectory that would target Luna 3 under the Moon's south pole, then use lunar gravity to loft it back over Earth's north pole. As Luna 3 approached the Earth, it became circumpolar, that is, not going below the horizon all day. Luna 3 was thus the first space probe to use the gravity of another celestial body to modify its path through space. The necessary trajectory was calculated by a team led by Dmitry Okhotsimsky at the Steklov Institute of Mathematics using a room-sized vacuum tube powered computer called the Strela-1. After being gravitationally deflected by the Moon, Luna 3's polar orbit around Earth ranged from 40,671 to 469,306 kilometers with a period of 15.3 days.

When Luna 3 passed the Moon, its slow spin was stopped by gas jets. A photoelectric sensor on the end opposite the cameras aligned toward the Sun. Control of the craft was then passed to a Moon-sensor on the camera end of the craft and the camera porthole covers opened automatically when photography began. Photography was timed to occur when the probe was at a point between the Moon and the Sun while the Earth was off to one side, thus preventing Luna 3 from accidentally locking onto the Earth.

Photography began on October 7 while the probe was 65,200 kilometers from the Moon and continued for 40 minutes while Luna 3 receded another 3,200 kilometers from the Moon. Thirty percent of the Moon visible from Earth was also in the view as a reference. Luna 3 thus photographed 70 percent of the far side not visible from Earth. Unfortunately, the apparatus using the long focus lens jammed after a few frames.

Facsimile transmission of the images began the next day while Luna was still beyond the Moon. Jodrell Bank also tracked the probe on frequencies of 39.986 MHz and 183.6 MHz using positional data and transmission times supplied by the Soviets. As Luna made its closest approach, signals received by the British on 183.6 MHz changed in nature. For 20 minutes, they alternated between steady 15-second periods and 15-second gaps. This was followed by 80 minutes of "beep-beep-beep" pulses, then the tone returned to the original pattern. Unfortunately, graduate students operating the Mark 1 telescope receiver narrowed the bandwidth to improve the signal-to-noise ratio and in the process "clipped" the video information. The resulting image intercepted by the Mark 1 radio telescope was of such poor quality that its content could not be determined.

Because there were only a few watts of power allocated to the radio system, it was designed to transmit the images slowly at great distances and more quickly

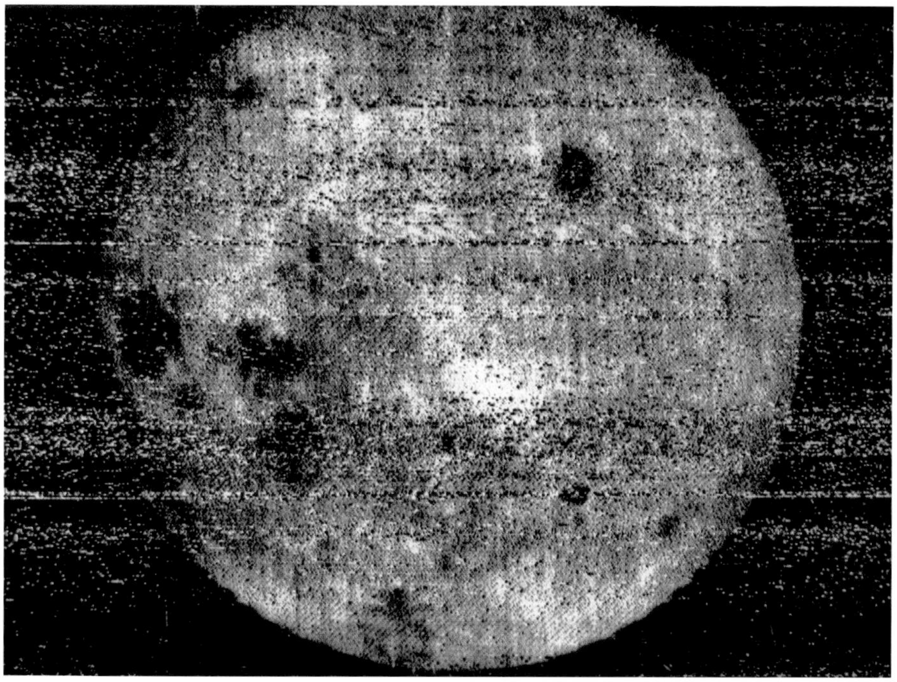

The original unprocessed farside images transmitted by Luna 3 were degraded by poor signal reception and radio static. Mare Crisium is visible on the left lunar limb and the dark spots of Mare Moscoviense, the Moscow Sea, and the Tsiolkovsky crater lie above and below center right.

closer to Earth. Image transmission began while still 470,000 kilometers from Earth. It took about 30 minutes to send a 1,000-line image. At translunar distances, the strength of the signal received from Luna 3 was about 100 million times less than that of ordinary analog television.

The first four attempts to transmit the photos at great distances were unsuccessful as radio static washed out the images. When printed on thermographic paper, all they showed was the Moon was round. On the fifth attempt, as Luna moved within 40,000 kilometers of Earth on October 18, two usable images were received. Of the 40 frames of film carried, 29 were exposed and 12 of those were transmitted to Earth. Eight hours were required to transmit all the images received, but the video transmitter only operated for an hour at a time with a two-hour break for recharging the batteries, thus 24 hours were needed to receive all the images.

The historic images were analyzed by Iu.N. Lipskii and his team at the Sternberg Astronomical Institute. Many of the images showed a distinctly elliptical Moon because the scanning mechanism on the ground failed to synchronize with the unit aboard the probe. Even on the better photos received, the images were fogged and streaked by radio interference caused by the rotation of the probe that varied the signal strength of the lobe-shaped antenna pattern sweeping past Earth. On October 22, both radio systems failed, preventing photo retransmission on succeeding orbits.

The lunar far side photos were a public sensation and graced the front pages of newspapers worldwide. A stunning two-page display in *Life Magazine* erased millennia of human ignorance about the far side of the Moon for millions of readers. However, the images were of poor quality. The need for wide area coverage on the first far side photo mission dictated the Moon be

Following the destruction of Atlas 9C during a ground test, Atlas 20D was substituted to launch Pioneer P-3 on November 26, 1959. The launch failed when the Atlas-Able nose cone ripped off the rocket one minute after launch.

fully illuminated, resulting in flat low-contrast lighting. The images had the same approximate quality as daguerreotypes taken of the near side of the Moon in the 1840s. After enhancement, gross features could be identified on 9 frames. The far side photograph, first appearing in the newspaper *Pravda*, then released to the world on October 27, 1959, was a composite of three separate photos.

The far side proved different than the near side in that only two small dark mare areas were visible. One, 300 kilometers in diameter, was named Mare Moscoviense (Moscow Sea) and the other was Mare Desiderii (Dream Sea). Later, the Dream Sea was renamed Mare Ingenii, or the Sea of Cleverness. What had been perceived on the initial images as the Soviet Mountains turned out to be a crater ray structure and the mountain designation was dropped. Other recognizable features were named after famous writers, scientists and philosophers such as Thomas Edison, Louis Pasteur and Konstantin Tsiolkovsky.

America and the USSR Both Stumble

American reaction to the latest Luna was markedly quieter than that which had greeted Luna 2 a month earlier. Soviet authorities continued to play up their latest achievement and the washed-out high-contrast image showed little detail other than the gross bright and dark features, but it nevertheless adorned hundreds of astronomy textbooks and space histories. Better images were created from magnetic tape recordings and released in 1961 as the *Atlas of the Far Side of the Moon*.

Two months after the Luna 3 spectacular, another NASA Atlas-Able was ready for flight, this time carrying the craft known as Pioneer P-3. Forty seconds after the November 26, 1959 launch, the lunar effort came to a sad end as the booster broke up and exploded. Analysis showed that the Atlas-Able's payload shroud had failed. To house the one-meter-diameter spacecraft and its folded solar panels atop the narrower Able stages, the shroud had a bulbous hammerhead shape. As the booster accelerated, the rush of air past the enlarged tip caused a partial vacuum on the outside of the shroud. The buildup of air pressure inside the shroud finally blew it apart and tore off the Pioneer spacecraft and its third stage.

Six months after the Luna 3 success, the Soviets again tried to launch two more 300-kilogram photographic lunar probes. These were to fly closer to the Moon and take higher resolution photographs of the western far side not seen by Luna 3. The Yenisey-2 photographic system was upgraded with a 750-mm focal length lens and the transmitter power was doubled to correct the photo transmission problems encountered by Luna 3.

The first flight on April 15, 1960, nearly succeeded. However, the third stage fuel tanks had not been fully fueled and the engine shut down three seconds early. An hour after launch, it was apparent that the photo probe was 130 meters per second short of the velocity needed to reach the Moon. The probe looped upward for 200,000 kilometers, then fell back to Earth in secrecy.

As soon as the misfire was recognized, the backup spacecraft was erected on the launch pad. Time was critical as the proper illumination of the Moon's far side would only occur on certain dates. Equally important was Khrushchev's desire for a lunar success before the May Day celebrations and the May 1960 international summit meeting in Paris. What happened the following day ranks as one of the most bizarre launches from Baikonur.

As the afternoon countdown reached zero, a group of observers, including Colonel Alexander Koshits, stood about 1.5 kilometers behind the launch pad. In his memoir, Colonel Koshits related how one of the first stage strap-on boosters failed to ignite and the SL-3 remained on the pad for an unusually long time. Once the booster slowly rose, it began to tilt over. At 150 to 200 meters in altitude, control was lost and all four strap-on boosters separated. Two fell near the launch pad and exploded. A third rumbled on toward Koshits and his fellow observers. The men jumped into a trench as the booster roared 30 to 40 meters over their heads. A cloud of sand blew over them just before the booster crashed and exploded some distance away. After the blast wave passed,

The launch of Ranger 3, America's first attempt to land on the Moon, was broadcast live on national television on January 26, 1962.

The USSR Also Plans a Lunar Landing

In response to the awakening American lunar landing program, the Soviets also set their sights on the same goal, though not driven by science alone but by the desire to demonstrate their country's technical superiority. Thus in 1961 and 1962, Korolev was pushed to design a craft capable of a lunar landing.

The same scientific debate about the condition of the Moon's surface also faced the Soviet design team. Was the lunar surface either hard rock or soft powdery dust? Sensing the impasse between the astronomers, geologists and engineers, Korolev stated categorically that the design should assume the lunar surface is hard. Legend has it that when challenged about who will vouch for that fact, Korolev shot back, "Me." Taking a piece of blank paper, Korolev wrote, "The Moon's surface is hard. (Signed) S. Korolev."

The second-generation Soviet lunar probes were to be rough landers, decelerating to a controlled crash onto what was anticipated to be a surface like stony beach sand. The evolving craft, known as the Ye-6, bore Korolev's trademark of simplicity in design to accomplish two goals, lunar landings and delivering satellites into lunar orbit.

The second-generation Lunas required only 3.5 days to reach the Moon and drew their power from batteries, simplifying the design of the 1,500-kilogram, 2.7-meter-long spacecraft. The bottom portion housed a throttleable KTDU-5A engine that was capable of up to 4,650 kilograms of thrust for 50 seconds. The engine, developed by Alexei M. Isayev, was remarkably advanced, featuring a turbopump-fed rocket with surface tension propellant management for zero-g ignition.

Above the engine was a one-meter diameter pressurized cylindrical compartment containing the control and communications equipment as well as orientation sensors. Mounted on either side of the upper fuel tank were two "saddlebag" equipment modules, each weighing 150 kilograms. One housed the radar altimeter that triggered the ignition of the retrorocket and the gas-powered attitude control jets. The other module contained the Sun and Moon sensors for attitude reference and most of the spacecraft batteries. Just before ignition of the retrorocket, these modules were jettisoned in order to lighten the spacecraft during powered descent.

For landing missions, the engine was budgeted a 100 meters per second velocity change for course corrections while the remaining 2,600 meters per second capability was used for landing. Lunar orbit missions required less deceleration to enter orbit, thus the tanks were only partially filled to allow about 930 meters per second braking into orbit.

The lunar landing payload was a pressurized 58-centimeter diameter 100-kilogram spherical capsule. The lander carried command and telemetry electronics, batteries, a television-like camera to relay a panoramic photograph of the surface, instruments to measure radiation and temperature levels and test the chemical makeup of the lunar soil.

The landing capsule was ejected from the descent stage just before touchdown. The heavy batteries and radio equipment were placed at the bottom to lower the capsule's center of gravity and help stabilize it in an upright position

The Soviet Luna Ye-6 lunar landing spacecraft stood 2.6 meters high and had a mass of 1,538 kilograms. The course-correction and lunar-braking rocket was housed in the conical tail, the central instrument section controlled the spacecraft, and the spherical landing capsule was carried at the top.

after landing. A water evaporation system maintained a temperature between 19 and 30 degrees Celsius.

A 1.5-kilogram, eight-centimeter-wide, cylinder-shaped scanning facsimile system popped up from the top of the lander and focused from 1.5 meters to infinity. Panoramic photos of the local terrain were taken from a height of 60 centimeters and the camera system was tilted 16 degrees varying the panoramic view from extreme close-ups out to the horizon. The rotating camera turret acted like a submarine periscope and used a nodding mirror to slowly scan 360 degrees in 100 minutes. The light beam from the moving mirror was focused by a lens onto a photoelectric detector. The signal was then amplified and transmitted to Earth. The total panorama was made of 6,000 vertical scans containing 500 pixels each. The system had a resolution of 3.6 arc minutes, or about one-third that of the human eye, allowing nearby objects as small as two millimeters to be imaged.

The SL-3 Luna booster which had sent Luna 1, 2 and 3 on their way was insufficient to loft the heavier landers. In the search for increased rocket weight-lifting capacity, Gleb Maksimov mathematically proved that the lunar payload could be increased from 400 kilograms to 1,800 kilograms by launching from an Earth parking orbit instead of using a direct ascent trajectory like previous Lunas. The new heavier probes would use the more powerful SL-6 Molniya booster. The proven R-7 rocket was retained as the first and second stages, but the Molniya used a fourth "escape" stage developed by Vasily Mishin. The new Luna landers were initially launched into an Earth parking orbit at about 200 kilometers of altitude. The SL-6 fourth stage and the Luna probe, weighing a combined 7.5 tons, would circle the globe for an hour before igniting as the craft crossed northward over the equator near South America. The escape stage would accelerate the Luna probe into a 3.5-day trajectory toward the Moon.

The guidance used to target the Luna landers was crude compared to American standards but retained the Korolev trademark of using the simplest system to maximize the chance of success. After several days of optical and radio tracking established the probe's trajectory, a two-dimensional course correction was calculated; that is, all course correction burns were done radially to the path of the trajectory. This corrected the aim point on the Moon's surface, but not the approach velocity. At this point, accuracy was essential. For every one meter per second error in the rocket burn, the probe would miss its intended target by an additional 1,000 kilometers.

Using this simplified guidance system, the Luna's approach velocity was not important, but the aim point was critical. The landers were targeted to a point between 62 and 64 degrees west along the lunar equator. By aiming at this limited locale with a 3.5-day trajectory, the motion of the Moon's orbit naturally intercepts a descending spacecraft on a line perpendicular to vertical. The craft's horizontal velocity is thus zero as it approaches the surface, eliminating the need to cancel horizontal motion, but limits the area on the Moon where a landing can occur.

During the final hour of flight, all the craft had to do was hold an attitude parallel to the trajectory for its engine thrust line to pass through the Moon's center. The craft would then be in the proper attitude when a radar altimeter triggered the retrorocket at about 75 kilometers of altitude. A computer throttled the engine in response to the altimeter to cut the velocity from 2.6 kilometers per second to under 10 meters per second. A proximity sensor mounted on an extended 5-meter arm detected the lunar surface and shutdown the rocket motor. At the same time, a powerful spring mechanism ejected the shock-resistant landing capsule away from the area disturbed by the retrorocket blast and clear of the rocket body in case the fuel tanks exploded upon impact.

Shortly after touchdown, a latch released the upper half of the lander's shell that split into four spring-loaded petals. When these opened, they forced the probe to remain in an upright position and acted as transmitting antennas. The imaging apparatus turret and four spring-loaded antenna rods for receiving commands from Earth then popped up.

After the difficulties receiving signals from the Luna 3 mission, it was realized that improved tracking and data acquisition facilities were needed not only for future lunar probes, but missions to the planets as well. A commission selected a location near Yevpatoria on the shore of the Black Sea as the site for the new deep space antenna capable of receiving signals from a craft up to 300 million kilometers away.

The engineering challenge of quickly building a 100-meter fully steerable antenna like the 76-meter Jodrell Bank dish proved too great. Ultimately, Yevgeni S. Gubenko of the SKK-567 Design Bureau created eight separate 16-meter dishes on a common mount. The massive antenna was known as the ADU-1000 and had a surface area of 1,000 square meters. Korolev himself came up with the design solution of how to mount and aim the separate dishes together. Taking note of the Jodrell Bank dish in England, Korolev also used a huge gun turret from a decommissioned battleship. In this case, the steel

To provide high resolution photographs of potential Apollo landing sites, NASA contracted the Boeing Aircraft Company to build the Lunar Orbiter satellites that utilized a photographic system originally designed for U.S. Air Force reconnaissance satellites.

response, the Lunar Orbiter project was started under the management of NASA's Langley Research Center. On May 10, 1964, Boeing Aircraft received an $80 million contract to build the Lunar Orbiter spacecraft. Kodak was subcontracted to provide the photographic system. The Lunar Orbiter program would provide 500,000 times the high-resolution photographic coverage of a Ranger impact mission.

In the meantime, the revitalized Ranger Block III series was prepared for photographic close-up reconnaissance of the Moon. By mid-summer 1963, a test craft had completed 13 separate 66-hour simulated lunar flights in JPL's new 7.6-meter thermal vacuum space chamber with only one failure. JPL was confident Ranger would not let America down again.

Retaining the same hexagon shape and parabolic communications antenna as its predecessors, the 364-kilogram Block III craft was now equipped with an aluminum frame, unsterilized electronic brain, backup timer, redundant attitude controls, larger 60-meter per second course correction motor, double-sized solar panels and a spare battery. The new craft's only cargo was a six-pack of high-resolution television cameras built by the Radio Corporation of America (RCA). The 173-kilogram television unit was mounted in a tapered 1.5-meter-high polished aluminum tower mounted atop the craft's hexagonal body. Two full-scan cameras returned 1,152-line images while four partial scan cameras

returned 300-line images. The cameras used 25- and 75-millimeter lenses to view different areas below the probe as it fell toward the Moon. The craft had even been stripped of all other science instruments to prevent them from affecting the success of the mission.

Rangers 6 through 9 were designed to deliberately crash into the Moon while six television cameras transmitted closeup photos of surface details.

The camera-carrying spacecraft had been conceived to provide data about the lunar surface so the landing gear for the Apollo Lunar Module (LM) could be optimized. If designed for the worst anticipated conditions, the gear could account for up to 50 percent of the LM's mass. If the Moon's surface was smooth and hard, the landing gear could account for only 5 percent of the LM's mass. At a time when Apollo was fighting for every kilogram to be sent to the Moon, the question carried enormous weight, both figuratively and literally. However, delays pushed the first Block III launch into late 1963 and the time to freeze the LM's design had already passed. Plans now called for all four Block III flights to be completed by July 1964, allowing time for quick modifications to the LM if the Ranger data so dictated. But this ambitious flight schedule eroded as pesky electronic bugs delayed the completion of the television subsystem. It was not until January 1964 that Ranger 6 stood ready to make a suicide plunge into the lunar surface for the sake of Apollo.

Lifting off from pad 12 at Cape Canaveral, Ranger 6 bored into an overcast sky on January 30, 1964. The Atlas-Agena performed flawlessly, sending the television-laden craft on its way to the Moon. The only anomaly came two minutes after launch, when a telemetry signal showed Ranger's television system had inexplicably turned on. Then, strangely, the system turned itself back off 67 seconds later.

After the Agena finished propelling Ranger away from Earth, the lunar probe obediently opened its solar panels and antenna, then settled into a cruise to the Moon. Telemetry showed all was well and offered no reason why the television system came to life during the launch. Everything continued to check out as a textbook-perfect flight as the next day Ranger 6 executed a mid-course maneuver to aim toward the Sea of Tranquility.

A Shocking Failure

On the morning of February 2, Ranger 6 and the American Moon program prepared to meet its fate. As the Moon's gravity accelerated the craft, JPL awaited the stream of lunar close-up images. Dropping past impact minus 18 minutes, Ranger dutifully began to warm up its collection of television cameras. At minus 13 minutes, the images would begin.

Then, at the appointed time, nothing. Monitors at JPL remained blank. No Moon pictures, no lunar surface rushing up to greet the scientists. Frenzied backup commands were sent to the spacecraft and ... still nothing. While Ranger's telemetry continued its strong signal, the video signal was not to be found. Repeated commands to spacecraft brought no response. At 4:24 a.m., Ranger 6, blind as a bat, slammed uselessly into the Moon.

It was a horribly ironic moment. Ranger 6 had been stripped of all science instruments and flown with the single-minded purpose of providing lunar surface imagery to support the design of the Apollo LM. Now in the last 15 minutes of what had been a beautiful flight, the television system failed on what was an otherwise fully functional spacecraft. To grind home the frustration, Apollo now had no need for the data Ranger was sent to collect. The design of the LM had already been frozen. Ranger could only confirm that the LM's design decision was a good one. As far as Apollo was concerned, Ranger 6 had been an exercise in pure futility.

Once again NASA, Congress and the public railed at a lunar embarrassment. The mission represented a $26 million loss (uninflated pre-Vietnam War dollars), and numerous investigations probed the failure. Even the fictional comic strip detective Dick Tracy got into the act. On the comic page, he traveled to the Moon in a magnetically levitated spaceship supplied by Diet Smith, his Howard Hughes-like corporate mentor. There, Tracy located Ranger 6's crumpled remains. The famed detective's conclusion as to why the mission embarrassed America: the cameras were never turned on.

The real-life investigations into why the failure occurred were more complex. In the coming months, the explanation finally surfaced. When the Atlas rocket's two outer booster engines dropped off two minutes into the flight, 67 kilograms of unburned kerosene was released and vaporized in the air. The fuel cloud was then ignited by the sustainer engine and created a fireball that momentarily

Ranger 7, launched by an Atlas-Agena B rocket, lifted off from Cape Canaveral, Florida on July 28, 1964, and was America's first successful lunar probe.

engulfed the entire rocket. The pressure of this flash forced plasma from the burning fuel vapor past the edges of the Agena stage's umbilical door and briefly shorted umbilical connector pins wired through the Agena and leading to Ranger's television system. The brief connection was enough to activate the television system at the wrong time and short out critical electronics.

In the USSR, the lunar chase fared little better than the American Ranger embarrassment. After repeated failures of the SL-6 escape stage to leave Earth orbit, Korolev ordered a redesign of the rocket. Kosmos 21, launched November 11, 1963, was both a test of the refitted booster and a lunar flyby mission. However, the spacecraft was not a Luna. It was instead a test of a Venus flyby craft designated the MV-2. The probe was to photograph the remaining unseen portion of the Moon's far side while testing the Venus probe. The plan was spoiled when Kosmos 21 remained stranded in orbit then fell to Earth three days later.

Four months later, the Soviets tried to fly an improved version of the Luna lander on March 21, 1964, and again on April 20. In both cases the SL-6 booster failed to achieve Earth orbit. A June 4 launch followed, this time with another test version of the MV-2 Venus probe. However, the new photographic mission to the Moon's far side never reached space.

In an October 1964 bloodless coup, Khrushchev was overthrown as the leader of the Soviet Union. Ironically, his removal from power occurred during one of his coveted space spectaculars, the flight of the three-man Voskhod 1. Khrushchev had become almost synonymous with Sputnik and Moon probes. His bluster after each Soviet space spectacular came to be an expected part of the Space Race. Admittedly, future Soviet space activity would seem a little bit duller without Khrushchev. While political propaganda was still a major objective of Soviet space missions, science was now a higher priority.

By now, the Ranger program seemed to have made all the mistakes possible. To allow extra time for review and modifications following the failure of Ranger 6, the next flight was deliberately delayed until the last minute prior to when the launch complex was taken over by the Mariner 3 and 4 Mars missions in late summer 1964. Perhaps Ranger 7 would finally receive a lucky roll of the dice.

By July 28, all was ready. The Atlas-Agena pushed forcefully into a clear Florida sky. When the booster engines dropped away, Ranger 7's telemetry showed no glitches. The craft had passed the barrier that killed its predecessor. The next day, a course correction targeted Ranger 7 toward its destination north of the Moon's Sea of Clouds. So far it was a textbook flight and cautiously optimistic JPL officials predicted a 50/50 chance for success.

American Lunar Success

Early on the morning of July 31, Ranger 7 entered the Moon's gravity and quickened its drop to the surface. On cue at 9:07 a.m., the two wide-angle television cameras warmed up followed by the four telephoto cameras at 9:10 a.m. Ninety seconds later, years of frustration and heartache in trying to reach the Moon were forgotten in one ecstatic moment when clear video signals were received at JPL's Goldstone tracking station.

The initial photos were sent at an altitude of 1,800 kilometers, 15 minutes before impact. They equaled the best telescopic views from Earth. Now, as Ranger continued to drill home toward the Moon, each minute revealed greater lunar detail than ever seen before. The probe dropped closer and closer, still snapping pictures. By six kilometers of altitude, the resolution was tens of meters. When the last frame was taken 0.2 seconds before impact at 530 meters, the resolution was less than a meter over an area about the size of a football field. Then ... *impact.*

Ranger 7 was a smashing success, punching into the Moon at 2,615 meters per second. Years of disappointment, technical setbacks, political browbeating and public humiliation were swept aside. America had at last successfully reached the Moon. Technicians, project managers, politicians and newsmen gathered to view the event at JPL and in Washington reacted in a thousand different ways. Some cheered and leapt for joy, others openly wept. Spacecraft controllers abandoned their consoles, throwing papers in the air in a mini ticker-tape parade of back slapping and hand shaking.

Though destroyed, Ranger 7 accomplished its mission. It showed the lunar plains were smooth enough for Apollo to land. It also showed an ever increasing numbers of smaller and smaller craters, so many, in fact, that no room existed

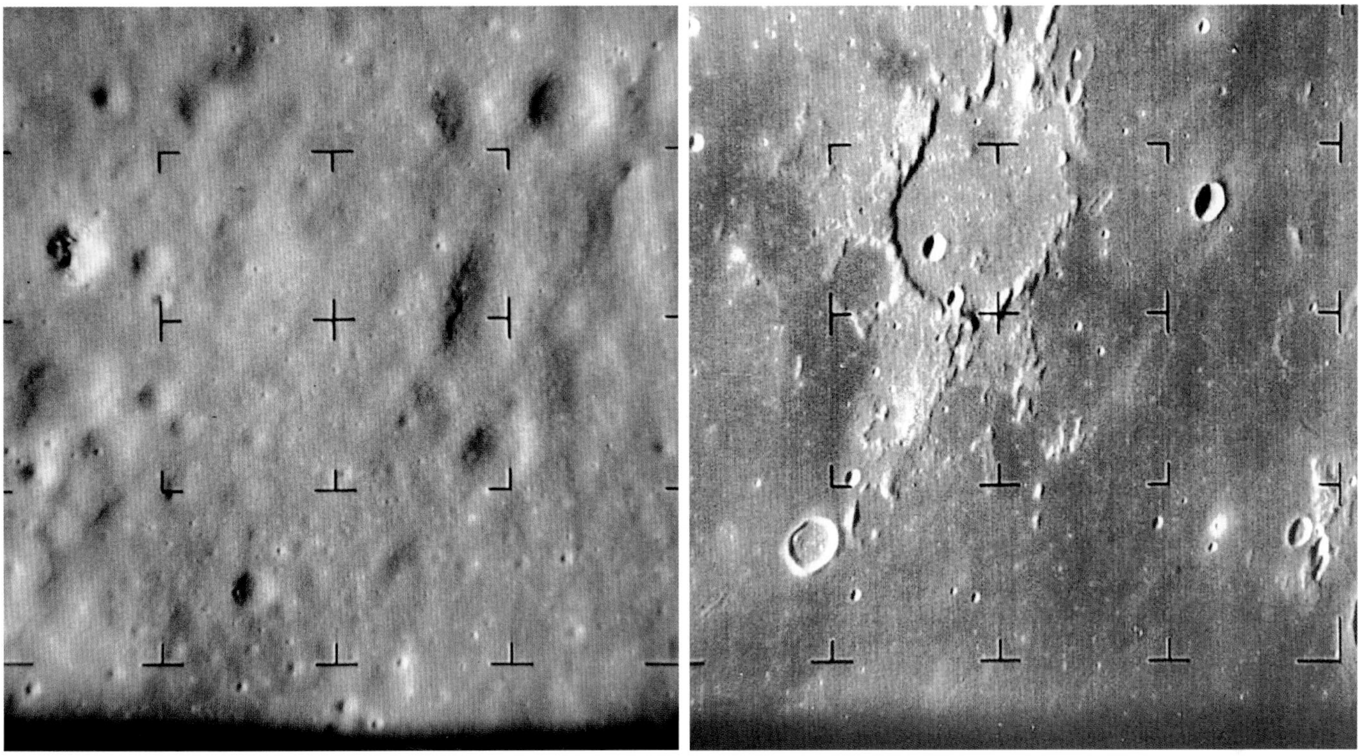

for additional craters. New craters overlaid old craters. New impacts filled in old craters and the net surface effect was a steady state of craters. The Ranger images showed the continuous meteoric impacts have produced a fine-to-coarse debris called regolith that covered the surface of the Moon.

During its plunge to the surface, Ranger 7 beamed back increasingly higher resolution images revealing never before seen details about the structure of the Moon.

Ranger 7 had not only impacted the Moon, it had impacted the public and lifted the spirits of America during a troubled time in its space program. Some believed Ranger's success was a greater achievement than sending men into orbit. Ranger 7 had redeemed America's Cold War humiliation from a time when the Soviet Sputniks danced in space while American rockets floundered. Newspapers from coast to coast carried multi-page displays showing a Moon that no human had ever seen before.

On a personal note, the American Moon triumph occurred just as I was in the middle of a military transfer from Travis AFB, located east of San Francisco, to San Antonio, Texas. As Ranger plunged into the Moon, my family was moving from base housing to an off-base motel in preparation for departure to Texas the next day. That night in the motel I watched a half-hour national television news special showing Ranger's magnificent images. With rapt attention, I saw the Moon revealed as never before. Later that night, I volunteered to sleep in the car because the motel room was very crowded. My secret agenda was I really wanted to devour the three San Francisco area newspapers and absorb the pages of lunar images and stories about the mission. Thus, like a child reading a comic book by flashlight under the bedcovers, I too soared to the Moon at midnight on July 31, 1964, from the parking lot of the Travis Sands Motel in Fairfield, California.

After the joy of the success subsided, scientists looked at what they had achieved. Ranger 7 had struck the Moon at 10.7 degrees south and 20.7 degrees west, only 16 kilometers from its aim point and sent back 4,308 photographs of the lunar surface. Astronomer Gerard Kuiper ecstatically proclaimed that Ranger had surpassed the resolution of Earth-based telescopes not by a factor of 10 as was hoped, but by a factor of hundreds. Kuiper concluded from the photos that the lunar surface was indeed strong enough to support the Apollo LM and that astronauts could (and would in 5 years) walk on the Moon.

In recognition of Ranger 7's achievement and to honor the thousands of people who struggled so long to reach the Moon, the area north of the Sea of Clouds near the crater Guericke where the craft impacted was renamed Mare Cognitum, or the Sea That Is Known.

As Ranger 8 was prepared to again reconnoiter the smooth lunar mare for Apollo, it almost seemed anticlimactic. The long uphill struggle to reach the Moon was over. The thrill of chasing the elusive target was gone. Now that the how of getting to the Moon was solved, the scientific investigation of the Moon could proceed.

Departing Earth on February 17, 1965, Ranger 8 flew another picture-perfect mission. Smacking the Moon only 24 kilometers from its aim point, the craft left a human-made crater at 2.6 degrees north and 24.8 degrees east. Returning 7,137 photos of the Sea of Tranquility. Ranger once again brought cheering engineers and congressmen alike to their feet. Perhaps now America's lunar curse really was over.

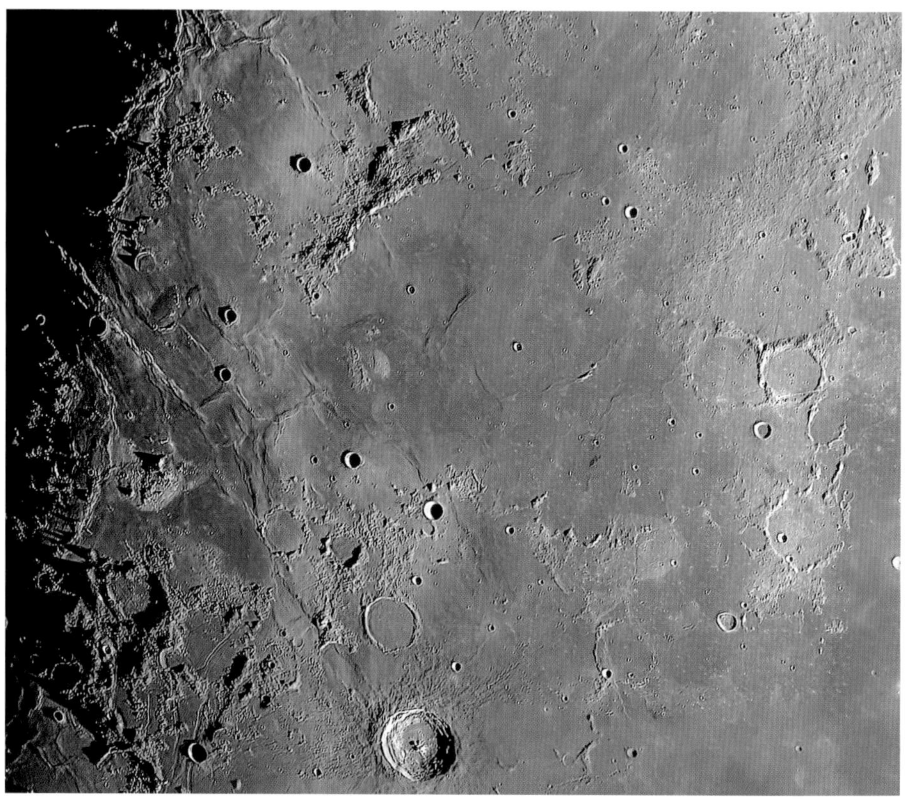

Ranger 7 impacted on what was then the southeastern portion of Oceanus Procellarum on July 31, 1964. The area bordered by the Riphaeus Mountains on the west and Guericke, the ruined crater with the crown-like rim on the east, was then renamed Mare Cognitum, the Sea That Is Known, in honor of the first successful American Moon probe.

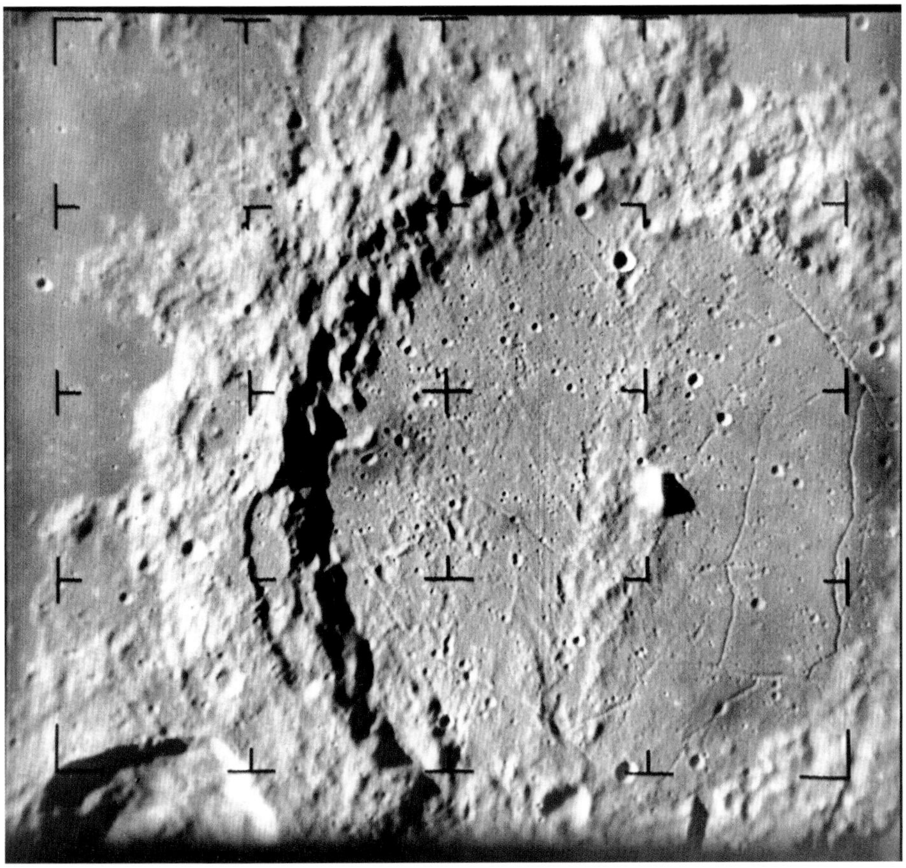

On the morning of March 24, 1965, a thrilled nationwide live television audience watched the crater Alphonsus loom ever larger as Ranger 9 broadcast live images of its approach to the Moon.

The new views were like those returned the previous July. But to some, they still didn't completely answer the question about whether the lunar surface could support the Apollo LM. As Thomas Gold, now with Cornell University lamented, the Ranger pictures were like mirrors ... every scientist saw his own theories reflected in them. Only the next step in lunar exploration, an actual landing on the Moon, would definitively answer the question.

An equally vexing question was who would achieve that landing first. America's unmanned Surveyor lander was nearly two years behind schedule and millions of dollars over budget. The Soviets were preparing a relentless drive to achieve their own landing at the earliest opportunity. Would America's newfound lunar luck hold out for Surveyor to make it to the Moon, or would a Soviet lander touch the surface first? Time would tell.

Concluding that Ranger could do no more for Apollo, the final television flight was directed toward a more scientifically interesting location than a flat target for manned landers. Ranger 9 thus streaked off toward the lunar highlands on March 21, 1965, and followed a course so accurate it would hit the Moon even without a correction. A day later, a brief rocket burn assured an impact would be within the 112-kilometer-wide crater Alphonsus.

The mission took on an added Cold War political urgency when just three days before, Soviet cosmonaut Alexei Leonov wiggled out of his Voskhod 2 spacecraft for the world's first spacewalk, an important step before sending men to the Moon. To counter the Soviet spacewalk propaganda coup, JPL engineers also had a marvelous public relations trick of their own. They had rushed completion of a device built for the Surveyor missions called an electronic scan converter. This device would allow Ranger's live photographs to be converted to the broadcast television signal standard. Now everyone could ride with Ranger to the lunar surface via a live nationwide television broadcast from the Moon.

At 8:48 a.m., March 24, Ranger 9's cameras began their warmup. Then, a magical scene never before encountered, from an altitude of 2,100 kilometers, the ringed crater Alphonsus slowly loomed ever closer as the entire nation experienced the thrill of live lunar exploration. At 9:08 a.m., Ranger crashed at 13.3 degrees south and 3.0 degrees west, near the 1,050-meter central peak within Alphonsus. The lunar picture show was over. In all, 5,814 increasingly detailed photos had been returned. Most of the public who watched the drama unfold really did not understand what they saw, but just being in a front row seat was worth the ride. Seven years later, a camera aboard Apollo 16 would locate the 14-meter diameter crater blasted by Ranger 9 as it struck the Moon less than seven kilometers from its aim point.

The Hard Road to the Lunar Surface

The first Soviet attempt to land instruments on the Moon during 1965 came just days before the success of Ranger 9. The effort was unsuccessful. The March 12 launch failed to leave orbit and fell back to Earth five days later. Two months later, on May 9, 1965, the 1,476-kilogram Luna 5 was to attempt a soft landing on the Moon. The latest Moon explorer was launched on the 20th anniversary of the World War II surrender of Nazi Germany and the launch announcement came just after a giant Victory in Europe Day military parade in Red Square.

On May 10, Luna 5 performed a mid-course correction burn that assured a lunar impact, but the craft was off course. Instead of targeting the western edge of the Moon, where the craft would descend perpendicular to the lunar surface, Luna 5 was headed for the Sea of Clouds, on the south side of the lunar equator. Luna 5 made no further corrections to restore its aim, indicating the course correction rocket likely failed.

At the 60-hour mark, the USSR broke the tradition of secrecy and announced the plan to soft land on the Moon. An official announcement said the mission's objectives were "to obtain the first experimental data on ... a soft landing on the Moon."

After 75 hours of flight, Luna 5 picked up speed due to lunar gravity. Then at the appointed altitude, nothing happened. Luna's retrorocket failed to fire, reinforcing the notion that the engine had malfunctioned during the previous course correction maneuver. Luna 5 crashed into the Sea of Clouds on May 12, five minutes before the anticipated touchdown time, at 31 degrees south and 8 degrees west. Even if the retrorocket had fired properly, the craft would have been destroyed because of excess horizontal velocity.

Only 30 days later, Soviet technicians tried again with the launch of Luna 6 on June 8, 1965. After a brief stay in an Earth parking orbit, Luna 6 embarked

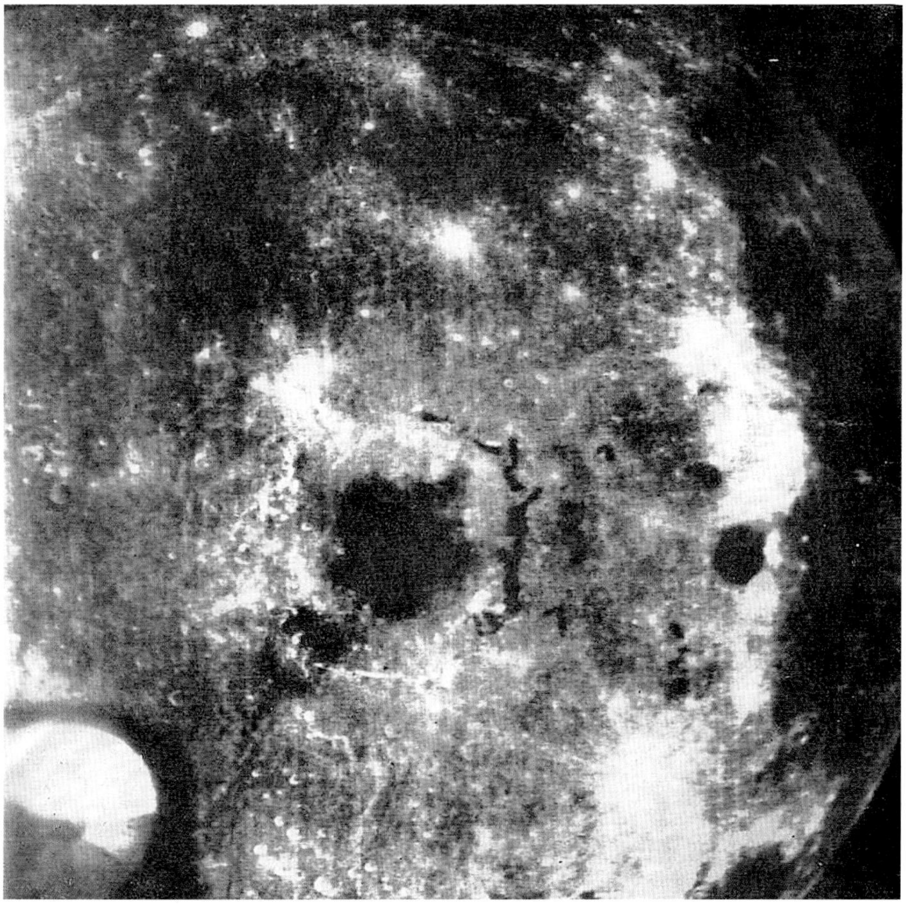

On July 20, 1965, the Soviet Zond 3 passed 9,200 kilometers from the Moon and photographed the lunar far side not seen by the previous Luna 3. The full extent of the basin structure surrounding Mare Orientale was revealed for the first time.

on another soft-landing attempt. The plan was thwarted, however, by a stuck engine controller which failed to shut off the mid-course correction rocket burn on June 9, 1965. By accidentally expending all its fuel, Luna 6 performed a 2,600 meter per second plane change and missed the Moon by a record 160,000 kilometers on June 11, 1965.

July of 1965 was a banner month in the fledgling international space exploration effort. On July 14 the American Mariner 4 craft had flown past Mars and showed it to be a cratered, lunar-like and apparently dead world. After the Luna 3 reconnaissance of the Moon's far side in 1959, there remained approximately 20 percent of the lunar surface that had not yet been seen by human eyes. This gap was filled July 20, 1965, when the Soviet Zond 3 Mars probe, launched two days earlier, photographed the rest of the lunar far side while passing the Moon on the way to the Red Planet's orbit.

The photographic system on Zond 3 was designed by Arnold Sergeevich Selivanov and was an improved version of the one used by Luna 3. Photography of the hidden side began as the craft approached within 9,220 kilometers of the Moon. A smaller portion of the Moon was photographed than on the previous Luna 3 mission, but lighting conditions were more favorable and surface relief was defined by long shadows. Twenty-five photographs were taken at intervals of 2 minutes and 15 seconds. Because photography extended over an hour, the probe's angle from the center of the Moon changed by 60 degrees during the flyby allowing stereoscopic study of the far side of the Moon.

Nine days after flyby, while more than two million kilometers from Earth, the far side photographs were transmitted, each taking 34 minutes to arrive. The nine day delay was necessary until the probe had receded far enough away to allow the directional high-gain centimeter-band antenna to be properly focused on Earth. The transmissions were repeated when the probe had receded to 31.5 million kilometers from Earth.

In September 1965, Soviet astronomer Alla Masevich revealed some preliminary studies of the Zond 3 photos where she noted the resolution was three kilometers and showed more than 1,000 previously uncharted craters. An interesting feature seen on the far side, but rarely on the near side, are chains of craters, some reaching up to 1,500 kilometers in length. Masevich also revealed that like the Luna 3 views, the Moon's far side is largely devoid of mare, is lighter in color and is more mountainous. However, there existed mare-like depressions the Soviets called "thalassoids" (from the Greek word "thalassa," or sea). Some extended up to 500 kilometers. These thalassoids were structurally different than nearside mare in that they lacked the familiar dark color. Today, these depressions are recognized as impact basins and the term "thalassoid" has been discontinued.

The Zond 3 images of the Mare Orientale west of Oceanus Procellarum were quite revealing. The circular structure of Orientale had been known since William Hartmann deduced its true nature while assembling Gerard Kuiper's famous *Rectified Lunar Atlas* in the early 1960s. Now another piece of lunar truth fell into place: chains of craters originating from basin impacts. The link between basins and rows of aligned craters was tenuous at the time. Now the Zond images of the huge Orientale bullseye left no doubt that the radial grooved sculptured terrain seen in some areas of the near side was the same type of feature seen clearly radiating out from the Orientale impact. Orientale, being one of the last basin impacts, had not been degraded by ejecta from subsequent impacts and clearly showed the link between the basin and chains of secondary craters. Following the lead from Orientale, investigators were now able to recognize many near-side craters as secondary craters associated with specific basin impacts.

Three months later, technicians completed modifications to the Luna design to hopefully eliminate the troubles that previously plagued the spacecraft. On October 4, 1965, the eighth anniversary of the launch of Sputnik 1, the heavier 1,506-kilogram Luna 7 was propelled into a translunar trajectory. The 2.6-meter telescope at the Crimean Astrophysical Observatory obtained multiple photographic exposures of Luna 7 as it receded from Earth. Using the photos and other tracking data, ground controllers guided Luna 7 through a mid-course correction on October 5, 1965. When Luna 7's braking rocket ignited on October 8, 1965, it was apparently too soon; the spacecraft stopped at too high an altitude. Desperately searching for the surface, the spacecraft exhausted its fuel then fell toward the Moon. After 86 hours and 8 minutes of flight, Luna 7 crashed onto the Moon in the Ocean of Storms near the crater Kepler.

Following the 75 kilometers per hour impact, the spacecraft sent signals for 3 more seconds, then went silent. Thirteen hours after touchdown, Soviet scientists admitted that Luna 7 had failed.

Launched on December 3, 1965, Luna 8 weighed a record 1,552 kilograms. Beginning with Luna 8, the Soviets shifted from a standard 65-degree inclination parking orbit to one inclined 52 degrees. The shallower inclination made more effective use of the Earth's rotational speed to assist in reaching Earth orbit and allowed the rocket to carry 46 kilograms of additional payload.

Two days after launch, Luna 8 performed a successful course correction that targeted it toward the Ocean of Storms. Soviet scientists apparently had high confidence that this mission would succeed as they announced Luna 8 would study the load-bearing qualities of the lunar surface.

All went well until the very end, when the craft's retrorocket fired several seconds late. Even at full thrust, Luna 8 could not overcome its own inertia, and after an 83-hour flight it crashed on the Ocean of Storms on December 7, 1965, near the crater Galilei. For a time, there were fears that Luna 8 had successfully landed but had been swallowed up by soft lunar dust.

New Leadership for Soviet Efforts

In 1965 all unmanned lunar and planetary programs, including the Ye series of lunar probes, were transferred from Korolev's control to the S.A. Lavochkin State Union Machine Building Plant led by Georgy N. Babakin. The transfer freed Korolev to devote more resources to the growing Soviet manned spaceflight program. Babakin himself was a "hands-on" engineer who was basically self-taught and did not receive a college degree until age 43, just seven years before assuming the Chief Designer's position at Lavochkin. Babakin in fact harbored a certain dislike for formal education but was nonetheless a very gifted researcher. For 15 years, Babakin worked on guided missiles, particularly the huge Burya cruise missile. By 1960 he was deputy chief designer of guidance systems and rose to the head position at Lavochkin at age 50, just as Korolev transferred all automated deep space probes to Lavochkin in 1965.

Luna 8 would be the last mission that Chief Designer Korolev would oversee. He entered a hospital for medical tests in mid-December. In mid-January, he underwent surgery for what was believed to be hemorrhoidal bleeding. On the operating table, the surgeon found the real cause of Korolev's health problems: a large intestinal tumor. The doctor decided to perform the cancer surgery on the spot, but the operation went badly, and Korolev died on the operating table from massive hemorrhaging. He never lived to see his spacecraft design finally reach the Moon intact on its next mission only two weeks later.

With five lunar launches in 1965 alone, the Luna program was the most ambitious flight schedule yet attempted by Soviet space scientists. It was also clear that the program was facing the same technical crisis that haunted the early American attempts. The Soviet Union now held the lead not only in space firsts, but space failures as well. It was clear that Moscow was committing enormous resources in a furious attempt to be the first to land on the Moon before the

After the 1966 death of Sergei Korolev, the development of Soviet lunar and planetary spacecraft was passed to the Lavochkin Design Bureau headed by Georgy Babakin.

The spherical 99-kilogram landing capsule of Luna 9, the first spacecraft to land on the Moon, was designed to righten itself by opening four spring-loaded instrument covers, then use a slow scan camera on top of the lander to relay photos of the lunar surface.

Americans. But, safely hidden from world view by the cloak of official secrecy, Soviet engineers did not have to suffer the glare of world publicity when one of their Moon rockets went askew. When the American Ranger craft experienced four very public failures in a row, Oleg Ivanovsky, a Luna spacecraft engineer, noted, "We called the Rangers 'American kamikazes.'" But Ivanovsky knew well that the Russians were also having their share of growing pains in their effort to perfect a lunar landing spacecraft.

Lunas 4 through 8 all failed to achieve their intended lunar landing. However, as engineering exercises, all the failed flights were of great value in learning the technique of reaching the Moon. It was just a matter of time before Soviet science teamed with a bit of luck and landed on the surface of the Moon.

As the 1966 New Year arrived, both the Americans and the Soviets had hit the Moon. The dazzling close-up photographs returned by Ranger had given the public something more tangible than Luna 3's blurry and indistinct far side photos, and America was tall in the saddle again. But a critical Space Age goal was still up for grabs. A probe had to soft land on the Moon and report if a manned craft could survive. America's unmanned lunar efforts now shifted to the Surveyor lander and Lunar Orbiter photo reconnaissance projects. However, both craft were still far from completion and their maiden flights were not scheduled until summer. In the meantime, at Baikonur, another Luna was readied.

Liftoff came on January 31, 1966, and Luna 9 was on its way. On February 1, the craft performed a course correction to ensure accurate lunar impact. At

8,300 kilometers Luna 9 assumed a vertical attitude for landing. At this point, lunar gravity was accelerating the spacecraft by 1.3 meters per second. As the craft dropped past 75 kilometers, a radio altimeter fired the retrorocket. Jodrell Bank monitored Luna 9's signals as it approached its destination. The engine burned for 46 seconds while slowing the lander to 25 kilometers per hour. Touchdown occurred on February 3, 1966.

Luna 9 landed just after local sunrise at 7.8 degrees north, 64.22 degrees west in the Moon's Ocean of Storms. Signals from the Luna's descent stage died when it crashed. The television carrying landing capsule was thrown clear a second before impact. It bounced to the surface and rolled upright like a child's bottom-heavy toy, coming to rest halfway up the inner slope of a small 25-meter diameter crater. Scientists nervously passed the 4 minutes and 10 seconds it took for the capsule's stabilization petals to open and the antennas to deploy. Then success! A signal at 183.538 MHz established radio contact with the first craft to report from another world. The Soviets had finally done it. The craft they called an "automatic lunar station" had reached the Moon's surface. Now the question on anyone's mind was what was the Moon really like? The answer would have to wait another seven hours, until the Sun was high enough for photography to begin.

The historic initial photos from the Moon's surface were sent on February 4. A second panorama was transmitted later that day and a third was received the next day. Between the panoramas, the Sun elevation angles increased from seven to 14, then to 27 degrees, causing the shadow relief to change. The descent stage apparently crashed over the edge of a crater and was not visible. When Luna 9 was active, live signals from the lander were played over Moscow radio and television. The signals sounded like the slow ticking of a grandfather clock through the hiss of radio static.

Professor A.I. Lebedinsky reported that between the first and second panoramas, the lander shifted slightly as it settled into the lunar soil, then shifted further between the second and third panorama. The camera's perspective was thus moved approximately 100 millimeters, or about the distance between two human eyes. This fortuitous accident allowed stereoscopic study of areas around the lander.

Moscow released the Luna 9 panoramas several days after the landing. Academician Mstislav Keldysh said: "We can see the lunar surface consists of adequately durable rocks of the pumice or slag type. It turned out to be hard enough for the station not to sink." Academician Alexander Vinogradov announced that Luna 9 found no loose dust on the landing site. The views showed the lunar surface to be rough textured and of volcanic origin, covered with rocks and boulders of various sizes, but smooth enough to be easily traversed by a visiting astronaut.

The photos showed there were many overlapping depressions a few centimeters across, each of which cast a small shadow. The nearby rocks were sharply outlined monolithic structures which barely depressed the ground below them, indicating they were likely ejecta from a meteor impact. The craft did not electrostatically attract dust and there was no dust build up on the camera lens. In his interpretation of the photographs, Soviet selenologist Nikolai P. Barabashov reported "The upper layer of lunar soil is a sponge-like rough textured mass, scattered with individual sharp-edged fragments of various sizes. This layer is clearly strong enough to support heavier objects."

Scientifically, Luna 9's single most important discovery was the fact that Luna 9 rested firmly on a solid surface. This was welcome news to planners in the American Apollo manned Moon landing project. Luna 9's only other instrument, a radiation detector, showed the average radiation dosage on the lunar surface was 30 millirads per day, a level safe for human exploration of the Moon. It was deduced that this radiation was primarily from cosmic rays.

Space Piracy!

The historic first photos from the surface of another world also became embroiled in international intrigue when bureaucratic inefficiencies prevented the Soviets from expeditiously releasing the lunar photos. The incident began about three hours after moonrise on February 3, 1966, when Jodrell Bank began to receive six-minute bursts of signals from Luna 9. After a few minutes, Reginald Lascelles, the public affairs officer at Jodrell Bank, recognized the signals were not encoded, but had the same pattern as photo facsimile signals previously relayed from America via the Echo II communications satellite. On a hunch, the Jodrell Bank team sent a car to the *Daily Express* newspaper in Manchester to pick up a Creed-type wire service facsimile machine used to transmit news service photographs. The Luna signals were then fed into the newspaper facsimile machine.

Initially, there were problems in adjusting the facsimile scan rate. Then the correct combination was hit and incredibly, when the Luna video signals were fed into the machine, it began to produce a photograph of the lunar surface. By a stroke of insight and resourcefulness, the Jodrell Bank staff and assembled reporters were thus the first people outside the Luna control center in the USSR to view the stark landscape of the Moon. When Jodrell Bank director Lovell saw the resulting photograph, he could only utter, "Amazing!"

Within minutes, the photos were relayed around the world. The Soviets were outraged by what they called "piracy" by the British for having scooped them

On February 3, 1966, Luna 9 transmitted the first of three panoramas showing the Moon's surface was hard enough to support a spacecraft. The lander's tilted position allowed viewing nearby features a few millimeters in size as well as the distant horizon.

with the first release of the lunar images. In a TASS news report, Academician Anatoly Blagonravov verbally blasted Lovell for releasing the photos without obtaining "the information necessary for correct reproduction of the image." A Jodrell Bank spokesman said the photographs were of such international importance as to justify their immediate release.

What the team at Jodrell Bank did not realize was the video signals from Luna 9 were not synchronized to the same standard as contemporary international facsimile transmissions. Normally, the black to white transitions in facsimile image transmission corresponds to a change in audio frequency between 1.5 to 2.3 KHz. The Luna 9 transmissions used a frequency change of 1.2 to 2.0 KHz with a 1.1 KHz synchronization signal denoting the beginning of each new scan line. The difference in transmission standards was not enough to make the signal incompatible with wire services facsimile machines, but it was enough to alter the vertical perspective by a factor of between 2 to 2.5 times.

Not knowing the true scale of the lunar views as transmitted by Luna 9's photo apparatus, the British versions of Luna's surface images had the effect of making all lunar features appear taller and more jagged than they really were. The vertically exaggerated views had a familiar look by mimicking the lunar surface depicted almost a century before by Nasmyth and Carpenter. To the generations of lunar observers preconditioned to this exaggerated notion of the lunar surface, the distorted Luna 9 images had the feel of legitimacy.

While the unauthorized release of the distorted Luna photos by the British annoyed the Soviets, it did give world-wide credibility to their achievement at a time when many doubted they had really succeeded in landing on the Moon. It was several days later that Moscow finally released their own properly scaled version of the images.

Soviet engineers had soft landed Luna 9 on the Moon before NASA could do so and secured the Soviet Union another propaganda win in the superpower race to the Moon. Although national pride was still tarnished, NASA engineers could breathe a sigh of relief that the Gold Dust phantom was not going to slow down the Apollo program. Seven and a half years after both the Americans and the Soviets began their reach for the Moon, the single most important question facing future lunar efforts had been answered. The Moon's surface was hard.

4

Robots, and Men, Explore the Moon

A common Soviet practice in the early days of the Space Age was to attempt a Khrushchev-style "space spectacular" to coincide with a politically important event. In the spring of 1966, the 23rd Congress of the Communist Party was to be held in Moscow. It would be hard to beat the February landing on the Moon for sheer publicity value, but the Soviets could still pull off another lunar first if they could place a satellite in orbit around the Moon before NASA's first Lunar Orbiter mission flew in the summer.

Georgy Babakin proposed that a variant of his Ye-6 lunar lander, the Ye-6S, could be quickly built and lobbed into lunar orbit to coincide with the political events in Moscow. The lunar satellite became a crash project and was designed and built in 30 days and a launch was attempted on March 1, 1966. However, after the rocket's guidance system malfunctioned and stranded it in Earth orbit, the failure was designated Cosmos 111.

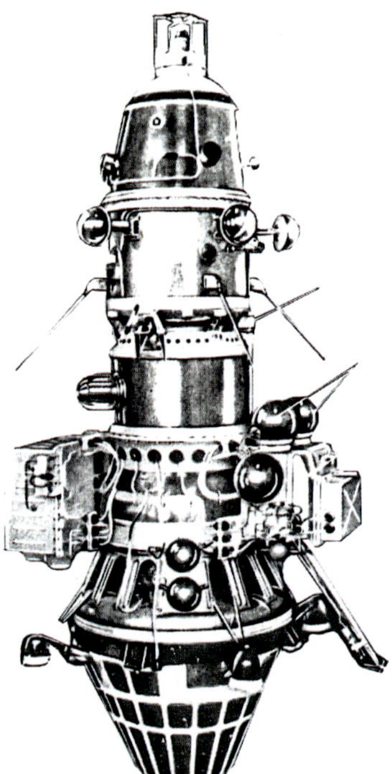

Luna 10, the first lunar satellite, was created by utilizing the existing first generation lunar landing spacecraft bus and substituting a satellite for the landing capsule. Upon arrival at the Moon, the spacecraft's retrorocket decelerated the satellite to allow it to be captured into lunar orbit on April 3, 1966.

A more successful shot lifted off from Baikonur on March 31. Official Soviet announcements stated Luna 10 was "to test a system ensuring the setting up of an artificial Moon satellite." Luna 10 had a mass of 1,582 kilograms and relied on the same spacecraft bus and propulsion system used on Luna 4 through 9. After an 80-hour flight, Luna 10 made its closest approach to the Moon and, on command from Earth, fired its retrorocket on April 3. The burn slowed the craft by 850 meters per second, allowing the Moon's gravity to capture the 245-kilogram Luna 10 satellite into a 349-by-1,017-kilometer polar orbit that looped the Moon every three hours.

The satellite was an airtight tapered cylinder 75 centimeters in diameter and 1.5 meters long with an extended 1.5-meter magnetometer boom. Instruments included micrometeor detectors, a gamma ray spectrometer, infrared detector, cosmic ray detector, X-ray, solar wind detectors and a three-axis magnetometer. Luna 10 was battery powered and thus had a limited lifetime of only 56 days.

Luna 10 arrived at the Moon during the 23rd Communist Party Congress meeting in Moscow. Not missing a major propaganda opportunity, the designers of the craft included a series of solid-state oscillators programmed to replay the musical tones of the "The Internationale," the international Communist anthem. Once in lunar orbit, one of Luna 10's first jobs was to bring the 6,000 party delegates to their feet by broadcasting the political anthem.

The first lunar satellite provided clues about the geochemistry of the Moon by studying gamma and X-ray radiation from the surface induced by cosmic and solar radiation. Luna 10's measurements implied the lunar surface rocks were roughly the same as basalt rocks on the Earth and that there was no significant difference in readings above the mare lowlands and the mountainous highlands.

Luna 10's magnetometer detected that the Moon had a weak magnetic field only 0.05 to 0.1 percent that of Earth's. The magnetic intensity did not change with increasing distance, leading researchers to believe that the observed magnetic field was merely the interplanetary magnetic field deformed around the Moon. They also concluded the Moon has no magnetic poles. The solar wind warps and stretches Earth's magnetic field away from the Sun, creating a long teardrop-shaped magnetic tail. Luna's magnetometer detected Earth's magnetic tail, establishing that it extends out to the Moon's orbit. Between April 3 and May 12, 1966, the meteoroid detector identified 198 impacts. The impacting particles were thought to be grain-like material blasted into lunar orbit by meteorite impacts.

The American Explorer 35 satellite that entered orbit around the Moon a year later disputed Luna 10's findings. It detected no lunar magnetic field and showed the solar wind flows past the Moon undisturbed.

Orbital tracking of Luna 10 provided information on strength and variation of the lunar gravitational field. These studies disputed previous ideas that the Moon was a 3,474-kilometer diameter sphere with a slight bulge toward Earth. Tracking the lunar satellite implied the Moon was pear-shaped with an elongation facing away from Earth. Repeated radio occultation experiments failed to detect any tenuous lunar atmosphere.

A month after Luna 10 departed for the Moon, an April 30 launch carrying another Soviet lunar orbiter failed to reach Earth orbit.

American Lunar Lander Struggles

In the early 1960s, the JPL-managed Surveyor program was envisioned as a $58 million project to reach the Moon by 1963. Unlike the Ranger, hard landers which were to splat onto the Moon, Surveyor was to land gently and utilize an array of instruments to investigate the Moon's surface. After a tortured six-year evolution, its cost had ballooned eight times to $469 million (1967 dollars). The project had been delayed for years not only by the complexities of developing the spacecraft itself, but also the equally vexing perfection of the hydrogen-powered Centaur upper stage booster needed to hurl the heavy spacecraft toward the Moon. But in early 1966, the spacecraft, the Centaur booster, NASA, and JPL were finally ready. Surveyor now represented the pinnacle of American robotic space exploration.

The American lunar lander used a combination of solid-fuel and liquid-fuel engines to autonomously approach the Moon and slow down to a feather-light touchdown. Designed as a 1,040-kilogram spacecraft, Surveyor stood 3 meters

The Lunar Orbiter Project Evolves

The Ranger and Surveyor projects existed prior to President Kennedy committing Apollo to land a man on the Moon. Surveyor was designed to autonomously land "blind" wherever its trajectory took it. But no one was willing to take such a chance with the Apollo manned landings. A means was thus needed to chart the Moon at high resolution to search for safe Apollo landing sites. Since the Surveyor orbiter had been canceled due to escalating program costs, a new lunar mapping satellite was proposed.

Despite the Boeing Aircraft Company's relatively high bid, the company won the contract and deployed a team of 1,700 people to build the new lunar mapping spacecraft. The project progressed smoothly from inception to orbital flight three years later. The only negative to the project seemed to be the public perception of its no-nonsense bland name — Lunar Orbiter. Despite pleas for a more imaginative name, NASA stuck with the craft's dull moniker.

The Lunar Orbiter photo mapping satellite evolved into a 387-kilogram spacecraft consisting of photographic and propulsion modules attached to a two-meter-high framework. Four solar panels spanned 3.8 meters and a 92-centimeter diameter steerable high-gain dish antenna and a boom-mounted

The Lunar Orbiter spacecraft was developed to photograph candidate Apollo landing sites. Designed by Boeing Aircraft and utilizing a photographic system previously created by Eastman Kodak for the Air Force Samos spy satellite, Lunar Orbiter could resolve surface features several meters in diameter.

The Lunar Orbiter photographic system was originally developed by Eastman Kodak for the secret Air Force Samos spy satellite program. The system utilized an 80-millimeter f/4.5 wide angle lens and a 610-millimeter f/5.6 telephoto lens. The photographic film was developed aboard the spacecraft and the images electronically scanned for transmission to Earth.

low-gain antenna on the opposite side spanned 5.2 meters. To slow the spacecraft enough for it to enter lunar orbit, the spacecraft used a propulsion system based on a 45-kilogram thrust Apollo spacecraft attitude control thruster. The rocket assembly was attached above the main spacecraft body and fueled by four tanks carrying nitrogen tetroxide and hydrazine.

The spacecraft was stabilized by gas jets to point within 0.2 degrees of local vertical during each photographic pass. A 10-watt transmitter relayed data back to Earth and a 128-word memory computer stored commands allowing 16 hours of photography which was then reprogrammed for the next photography cycle. The craft's only non-photographic experiments were 20 small, pressurized containers acting as meteoroid detectors and a series of radiation dosimeters to monitor the film condition in the cameras.

The Lunar Orbiter's key component was a 68-kilogram photographic system created by the Eastman Kodak Company. The need to image objects as small as one meter dictated a high-resolution system using photographic film instead of a television system. Housed in a pressurized ellipsoidal aluminum shell, the camera system was based on a photo system developed for the secret Air Force Samos photo reconnaissance satellite that operated from 1960 to 1962. At perilune, the lowest point of a lunar orbit, an 80-millimeter f/4.5 lens provided 8-meter resolution frames covering 906 square kilometers. The 610-millimeter f/5.6 lens simultaneously produced 1-meter resolution frames covering 52 square kilometers. The twin lenses peered through a quartz window. Seventy-nine meters of film allowed 212 images to be taken. Exposures varied between 1/25th, 1/50th or 1/100th of a second, depending on lighting conditions. An

Lunar Orbiter 1 returned the first image of the Earth as seen from the Moon. The photo of the crescent Earth over the lunar horizon taken in August of 1966 was a media sensation.

image motion detector viewing through the long-focus lens moved the film during exposure to compensate for the orbiter's motion across the face of the Moon. To develop the photos in the weightlessness of lunar orbit, the exposed film was pressed against another filmstrip coated with a monochemical layer to develop the film like a Polaroid camera.

To transmit the film images to Earth, Lunar Orbiter used a facsimile scanning system developed by the CBS television network. Each negative took 43 minutes to scan with a 0.005-millimeter diameter light beam at 287 lines per millimeter on the original negative. As the light beam varied by encountering different densities on the film, it was measured by a photomultiplier tube and the signal beamed to Earth for image reconstruction.

Lunar Orbiter 1 roared up from Cape Canaveral aboard an Atlas-Agena D rocket on August 10, 1966. After a journey of 92 hours, Lunar Orbiter 1 fired its engine and settled into a 191-by-1,854-kilometer equatorial orbit and became the first American spacecraft to orbit the Moon. The orbiter's primary goal was to photograph nine prime and seven secondary possible Apollo landing sites near the equator and to return close-up images of the far side of the Moon.

Lunar Orbiter Maps the Moon

Initial tests of the photo system between August 18 and 20 revealed that the motion-compensating sensor that was supposed to track the rapidly moving lunar landscape had failed and the high-resolution images were smeared. A decision was made by the Lunar Orbiter photographic scientists to leave the spacecraft in a higher orbit, bypass the faulty sensor, and map large areas of the Moon at 30-meter resolution. High-resolution Apollo landing site images would have to wait for the next mission. Unfortunately, Boeing project managers, anxious to achieve the mission's primary goal, convinced NASA to lower the orbit as planned in hopes the higher apparent ground velocity would jar the motion-compensating sensor into action.

On August 21, without the knowledge of the photographic team, the perilune was lowered to 58 kilometers, then further reduced to 50 kilometers on the 25th. The medium resolution images were outstanding, but the motion-compensation device still failed, smearing the high-resolution images and disappointing scientists. One public relations success was the first picture taken of Earth above the limb of the Moon, snapped on August 23 without the need for the motion sensor.

Precise tracking of Lunar Orbiter's path around the Moon revealed irregularities in the orbit. Something was making Lunar Orbiter dip slightly, then sail back up again as it circled the Moon. This was the initial American clue to the Moon's irregular gravitational field. The Soviets had also noted similar unusual orbital deviations in the path of their Luna 10 orbiter. Such information was crucial for Apollo lest the gravity anomalies tug the manned craft off its planned landing course. The perturbations were eventually traced to concentrations of denser rock, now known as "mascons," under the Moon's lava-filled maria, a discovery that was a major contribution to lunar science.

On October 29, its mission over, Lunar Orbiter 1 was deliberately crashed on the far side of the Moon so it would not interfere with the upcoming Lunar Orbiter 2 mission. Curiously, the spacecraft had not received a single meteoroid hit during its 8-week lifetime. A similar detector in Earth orbit would have expected an impact every two weeks. In all, the craft mapped five million square kilometers of the Moon and accomplished 75 percent of its goals. Preliminary photo results showed large rocks on the surface, suggesting it was strong enough to support a manned spacecraft. This proved that the successful landing of Luna 9 and Surveyor 1 was not a fluke. Lunar Orbiter 1 further suggested the Ocean of Storms possessed 20 percent fewer craters than other maria and would be a good candidate site for a manned landing.

Now that Soviet and American probes were both on the surface and in orbit, the road to the Moon had been paved and the highway was open for traffic. However, the next Soviet step toward the Moon stumbled. A launch on April 30, 1966 was intended to loft another lunar satellite, but the booster failed and the satellite never reached orbit.

Next up for the Soviets was Luna 11, launched on August 24, 1966. The spacecraft entered lunar orbit on August 28 and was designed to photograph the Moon as it traveled in a 159-by-1,200-kilometer orbit. Initially, there was some mystery about the fate of Luna 11. Jodrell Bank reported the craft's signals stopped four minutes after retrofire. The next night as the Moon rose above the English horizon, Jodrell Bank received a new burst of signals from the craft and there was no doubt it was in lunar orbit. However, Luna 11 was in trouble. The satellite went into a spin and the television system could not operate properly.

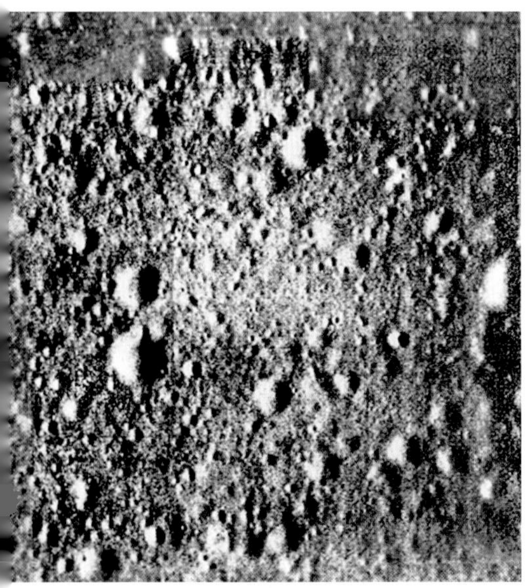

In October 1966, two months after NASA's Lunar Orbiter 1 returned high-resolution images of potential Apollo landing sites, the USSR's Luna 12 returned images of the Moon, although not as detailed as those from Lunar Orbiter.

Although photography was lost, Luna 11 repeated Luna 10's gamma and X-ray measurements to deduce the chemical makeup of the Moon's surface. The probe's instrumentation monitored the effects of giant solar flares which erupted on August 28 and September 2, 1966. Data was received for 277 orbits before the spacecraft's batteries were depleted on October 1.

Next up for the Americans was the Surveyor 2 liftoff on September 20, 1966. It carried only a television payload. Surveyor 1 had approached the Moon at an angle 6 degrees from vertical, but Surveyor 2 would approach from a riskier 23 degrees from vertical. At 16.5 hours after launch, a brief 9.8-second course correction burn by the three vernier engines was planned to target the probe toward Sinus Medii near the center of the Moon's visible disk. However, only two of the three engines ignited, pushing Surveyor 2 into a one-revolution-per-second tumble. Unable to properly direct the thrust of its retrorocket, Surveyor 2 crashed southeast of the Copernicus crater on September 23.

The Soviets finally mounted a lunar photo reconnaissance when Luna 12 was launched on October 22, 1966. The photo mapping system was housed in one of the strap-on modules attached to the spacecraft bus. On October 25, the Jodrell Bank radio observatory confirmed Luna 12 had entered lunar orbit. For photography, the equatorial orbit was lowered to 133-by-1,200 kilometers. On October 29, 1966, Moscow television showed two photographs of Mare Imbrium and the crater Aristarchus, taken from a distance of between 100 and 340 kilometers. Although they were not as sharp as those returned by NASA's

Lunar Orbiter 2 was deliberately turned on its side to image several targets on the horizon. The interior of the Copernicus crater as seen from a horizontal perspective was dubbed by the press as the "picture of the century."

Lunar Orbiter, they still showed objects as small as 15 meters. Spacecraft tracking confirmed the lunar gravity field was not uniform and variations in lunar gravity reduced Luna 12's perilune by three to four kilometers per day.

America's Lunar Orbiter 2 rose into the Florida sky on November 6, 1966, and followed its predecessor on a flawless flight to the Moon. The new orbiter's mission was to investigate 13 primary and 17 secondary Apollo landing sites north of the Moon's equator. On November 10 the Orbiter approached within 2,027 kilometers of the Moon and fired its braking engine to entered lunar orbit. On the 15th, the perilune was lowered to 50.5 kilometers for its photographic mission. By November 25, 208 photographs had been taken, including large portions of the far side of the Moon.

The Lunar Orbiter 2 mission is best remembered for the "picture of the century," an oblique view looking across the crater Copernicus from an altitude of 45 kilometers. The image was a sensation in the newspapers and showed the three-dimensional depth of Copernicus as the central peak rose a kilometer above the crater's floor. The orbiter also pinpointed Ranger 8's 12.5-meter-wide impact crater, revealing a raised central peak created by the upward rebound of bedrock after Ranger 8's impact.

On December 9, the orbiter's engine was fired for 62 seconds to change the orbital plane to 17.5 degrees in order to provide greater tracking data of the Moon's gravitational anomalies. During almost a year in orbit, Lunar Orbiter 2's micrometeorite detectors received three hits before the craft was deliberately crashed into the far side of the Moon on October 11, 1967.

The Surveyor 3 spacecraft was photographed by Apollo 12 astronaut Pete Conrad on November 20, 1969, two years and seven months after it landed on the Moon's Ocean of Storms in 1967.

what followed bordered on the bizarre. The craft touched down, then slowly rebounded 10.7 meters back into space, its engines still firing. A slow 0.3-meter-per-second horizontal velocity carried it 6 meters to the side, where it again rebounded 3.4 meters back into space and shuffled 2.4 meters further to the side. A third rebound took it back up one more meter before a command from Earth cut the engines and Surveyor 3 landed for the last time, sliding another 30 centimeters. The three-bounce episode took 34 seconds, and craft came to rest on a 14-degree slope.

Although the multiple touchdowns clouded the mirrors used to view under the craft, a study of the footprints provided extra data about the area's surface properties and proved conclusively that the Moon would support an Apollo lander. During its operation, Surveyor 3 returned 6,326 television pictures and performed a series of tests with the soil scoop. A total of 5,879 commands were sent over a total of 18.3 hours to dig four trenches, perform seven load bearing tests and 13 impact tests of the lunar soil. The surface material acted like coarse damp beach sand but could be easily dug.

Surveyor 3's electronics did not survive the frigid lunar night and contact could not be reestablished after sunrise. During its one lunar day of operation, Surveyor 3 sent back images of both Venus and the Earth as seen from the Moon. A total lunar eclipse on April 24 provided Surveyor 3 with the unusual perspective of seeing the Earth as a ring, with the Sun's illumination shining around the atmosphere.

An enduring myth is that the Surveyor 3 television camera recovered almost three years later by the Apollo 12 astronauts hosted Earth microorganisms that hitchhiked to the Moon and survived before returning to Earth. Post-mission analysis showed a colony of common bacteria, Streptococcus Mitis, had apparently stowed away in the Surveyor 3 television camera despite rigorous efforts to sterilize the spacecraft before its lunar journey. Scientists concluded that despite the vacuum conditions, near-cryogenic nighttime temperatures, and radiation on the Moon, 50 to 100 of the germs had survived the lunar round trip.

The claim about Surveyor 3's ultra-hardy Moon germs has never passed scientific peer review. A modern look at the biological "clean room" conditions where the Surveyor parts underwent examination reveals serious shortcomings. Far from the modern full-coverage biosuit used to isolate humans from examined samples, the Surveyor parts were handled by technicians wearing short-sleeve scrubs. Not only were their arms exposed, but the open scrub shirt tails were higher than the examining table. These conditions made it likely that the Surveyor 3 germs migrated to the television camera from the very technicians examining it after its return to Earth.

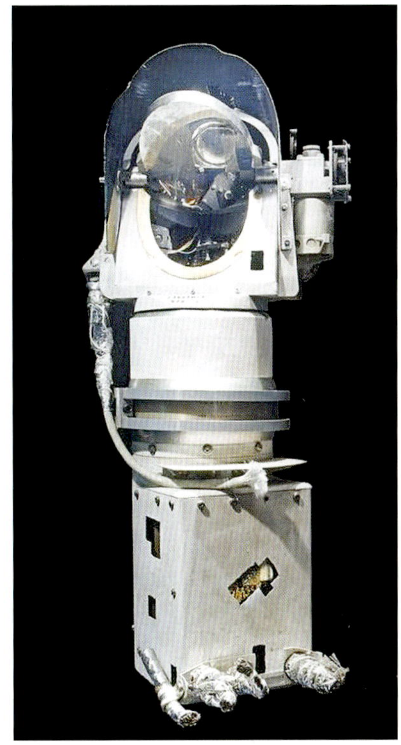

The Surveyor 3 camera system was removed from the spacecraft by Apollo 12 astronauts Pete Conrad and Alan Bean, then returned to Earth. The camera is now on display at the Smithsonian National Air and Space Museum in Washington, DC.

Science Becomes a Priority

With Apollo's landing site selection requirements fulfilled, the next Lunar Orbiter was dedicated to targets of scientific interest. After lifting off on May 4, 1967, the photographic probe continued the string of successes by becoming America's first lunar polar orbiter on May 8, circling in a high 2,705-by-6,034-kilometer path. Lunar Orbiter 4 mapped 99 percent of the near side at 60-meter resolution as well as part of the far side.

The understanding of the Moon's geology in the later part of the 20th century would not have been possible without the global lunar coverage provided by Lunar Orbiter 4. But the science mission was not without problems. A balky thermal door that covered the camera lens malfunctioned and threatened to blind an otherwise functioning spacecraft. The purpose of the thermal door was two-fold: keep the camera lens temperature stable to prevent it from fogging and protect the film supply from stray light that would degrade the images. Unlike the film in an enclosed camera body, part of the film inside Lunar Orbiter was uncovered as it fed through the camera and automatic developing mechanisms. Engineers got the thermal door open, and there it would have to stay. The trick now was to keep the lenses from fogging and prevent sunlight from striking the film looping inside the body of the orbiter. After noting that lens fogging was related to the incident Sun angle, mission scientist Howard Pohn came up with a solution to save Lunar Orbiter's precious few frames devoted purely to lunar science.

A low-tech way was used to visualize the spacecraft orientation as Lunar Orbiter soared over the Moon. Using a plastic coffee cup and some pencils to simulate the craft, engineers figured out how to maneuver Lunar Orbiter 4 so the Sun would warm the lenses yet not fog the film. The system worked and at an average cost of $100,000 per frame, $10 million worth of lunar data was saved. During subsequent photographic runs with the thermal door stuck open, only a single frame was struck by light. Thanks to the ingenious plan, Lunar Orbiter 4 would produce the grand image of the Orientale basin that confirmed the notion of impact origin for lunar basins as well as the first images of the Moon's south pole.

Surveyor 4 took off on July 14, 1967, and was again targeted toward Sinus Medii near the center of the Moon's visible disk. The spacecraft, a twin of Surveyor 3, was to dig into the lunar soil. A course correction brought the craft to within 3.8 kilometers of its aim point. On July 16, the lander's solid-fueled retrorocket fired and slowed the spacecraft. Then disaster struck. Two seconds before the motor was to burn out, contact was lost with Surveyor 4. It is not known if the motor exploded or if the craft silently completed the automated landing.

The sudden failure of an otherwise perfect mission was a shock to JPL and NASA. The previous craft sent toward Sinus Medii had also crashed and the destination seemed haunted. But Surveyor 4 would be NASA's last robotic failure on the Moon in the 20th century.

A satellite not usually associated with lunar exploration is Explorer 35. This craft was part of the Interplanetary Monitoring Platform (IMP) series that placed probes in deep space to study the solar magnetic field and solar particles in interplanetary space. Explorer 35 was a 104-kilogram flat octagon box-shaped probe with four paddle-wheel-style solar panels. It was launched from Cape Canaveral on July 19, 1967, by a Delta D booster. The probe entered a 764-by-7,886-kilometer retrograde orbit looping around the Moon every 12 hours. Once in lunar orbit, the craft was officially called AIMP 2, for Anchored Interplanetary Monitoring Platform. From its lunar vantage point, the probe studied magnetic fields, micrometeors, and the radiation environment of the Moon. A major finding was that the Moon does not have a magnetosphere like the Earth. Explorer 35 also performed the earliest bistatic radar experiment on the Moon by reflecting its 136.10-MHz telemetry radio transmissions off the lunar surface to the 45-meter Stanford University radio dish located in the California foothills between San Francisco and San Jose. After 2,167 days of

Surveyor 5 carried the box-shaped alpha scattering experiment that was lowered to the surface to analyze the chemical composition of the lunar soil. The device bombarded the soil with alpha particles emitted from a curium-252 source and measured the energies of the alpha particles reflected back to determine the chemical makeup of the soil.

operation, Explorer 35's experiments were turned off on June 24, 1973, when all of NASA underwent a tragic downsizing following the conclusion of the Apollo program.

Lunar Orbiter 5, launched on August 1, 1967, completed NASA's phenomenally successful series of Moon mapping spacecraft. The craft began photography on August 6 after its polar orbit was trimmed to 100-by-6,050 kilometers. By the 27th, all 212 photographic frames had been transmitted to Earth. The final orbiter covered 36 sites of scientific interest on the near side, five more potential Apollo sites, several Surveyor sites and completed mapping most of the far side. After the mapping mission was complete, the Apollo tracking network used Lunar Orbiter 5 to practice tracking a spacecraft in lunar orbit. Precision tracking allowed further investigation of the newly discovered mascons, showing they could indeed pull a manned Apollo lander off course.

The Lunar Orbiter missions returned 1,650 useful photographs out of a possible 2,210. But as successful as Lunar Orbiters were, the series could have been more productive, save for a last-minute penny-pinching decision by NASA. For only $13 million out of NASA's multibillion dollar budget, the already built Lunar Orbiter 6 could have been launched in 1968 to complete the exploration of the Moon's far side. Instead, the fully functional sixth orbiter was destined to blankly stare at tourists in Washington's National Air and Space Museum.

Surveyor 7 landed 29 kilometers north of the Tycho crater. Jet Propulsion Laboratory engineers assembled this astronaut's eye view mosaic of television images showing of some of the rubble hurled by the impact that created the Tycho crater about 108 million years ago.

The camera also detected two 1-watt lasers fired at the Moon from California and Arizona. Nighttime temperatures reaching -107 degrees Celsius damaged the spacecraft's battery and only limited contact was possible on the second lunar day.

A Lunar Scientific Treasure Trove

Overall, the Surveyor program had returned a scientific treasure of 87,700 television pictures showing the lunar surface, Earth, the Sun, bright planets and stars to the sixth magnitude. But more important for Apollo, the Surveyors had proven that the Moon was safe for a manned landing. Apollo could confidently alight on a surface made up of Earth-like basaltic material that behaved like wet beach sand.

However, while the robotic landings gave the blessing for men to follow the machines, the scientific results, although great, did little to resolve the riddle of how the Moon was formed. In the scientific "big picture," not much lunar science was returned by the Ranger spacecraft series, but the success of Ranger 7 was a tremendous morale booster after seven years of American failure to reach the Moon. The Surveyor spacecraft series did provide preliminary data on the composition of the Moon, but paradoxically, the program provided less support for Apollo than intended because it was conducted in parallel with the manned program instead of preceding it. All the Surveyor missions' lunar results could do is confirm that existing Apollo engineering decisions were, fortunately, correct.

Scientifically, the Lunar Orbiter missions were the most productive of the robotic explorers that preceded Apollo to the Moon. Lunar Orbiter drove the nails into the coffin of the volcanic crater formation theory. The long overdue burial of this idea finally allowed coherent investigations without a constant volcanic/impact origin debate clouding the science. Infrared studies from Earth showed hot spots on the Moon that some researchers interpreted as the remains of volcanic heat. Craters with smooth rims and high floors like the twin craters Sabine and Ritter had long been suspected of being volcanic. The Ranger 8 flight had been specifically flown over these craters to address the issue, but it remained just as murky after Ranger slammed into the Sea of Tranquility. Another suspect was the floor-fractured crater Vitello, also a well-known infrared hot spot. A Lunar orbiter 5 image showed the fractures on the crater floor contained large amounts of blocky material. Other fields of block-like rocks found by Lunar Orbiter also coincided with infrared hot spots. The suspected volcanic heat was nothing more than the blocks of rock that quickly radiated heat after sundown.

Circular dark halo craters were also long believed to be volcanic suspects. A Lunar Orbiter 5 frame showed the dark halo crater Copernicus H surrounded by more blocks of material. The dark halo craters were dark because they excavated dark mare material buried under the brighter ray material thrown from the Copernicus crater.

Another long-term misconception cleared up by the science-oriented Lunar Orbiter 5 was the issue of non-radial crater chains, that is, satellite craters that are not aligned radially away from the primary crater. Lunar Orbiter 5 zeroed in on the S-shaped stream of craters that extended north of the Stadius crater which lay in the ejecta field of Copernicus. A long line of respected researchers dating from the 19th-century pioneer Grove Karl Gilbert to geologist Jack Schmitt, who would himself later fly to the Moon aboard Apollo 17, all thought this crater chain was a collection of maar craters (craters created by volcanic steam explosions) overlaying the field of Copernicus' secondary craters. The high-resolution Lunar Orbiter frames quickly settled the issue by showing that members of the crater chain had the classic V-shaped herringbone secondary crater pattern that pointed back to the primary crater, Copernicus.

But Lunar Orbiter's news was not all anti-volcanic. More lunar clarity provided by the fifth Orbiter was establishing the true volcanic origin of features in the Imbrium area. The volcanic flow lobes on Imbrium are evident from Earth, but Lunar Orbiter 5 showed conclusively the pyroclastic dark blankets around Schroeter's Valley, Rima Prinz and Rima Aristarchus. The Marius Hills were also firmly identified as volcanic features.

A major gift from Lunar Orbiter was an understanding of the effect the impact basin-forming process has on the rest of the Moon. Between the great basins and moderate size craters is what can be called "middle scale" terrain. This includes radial ridges and grooves, dunes, plains, and crater clusters. These features are seen in abundance through the telescope, but how they were

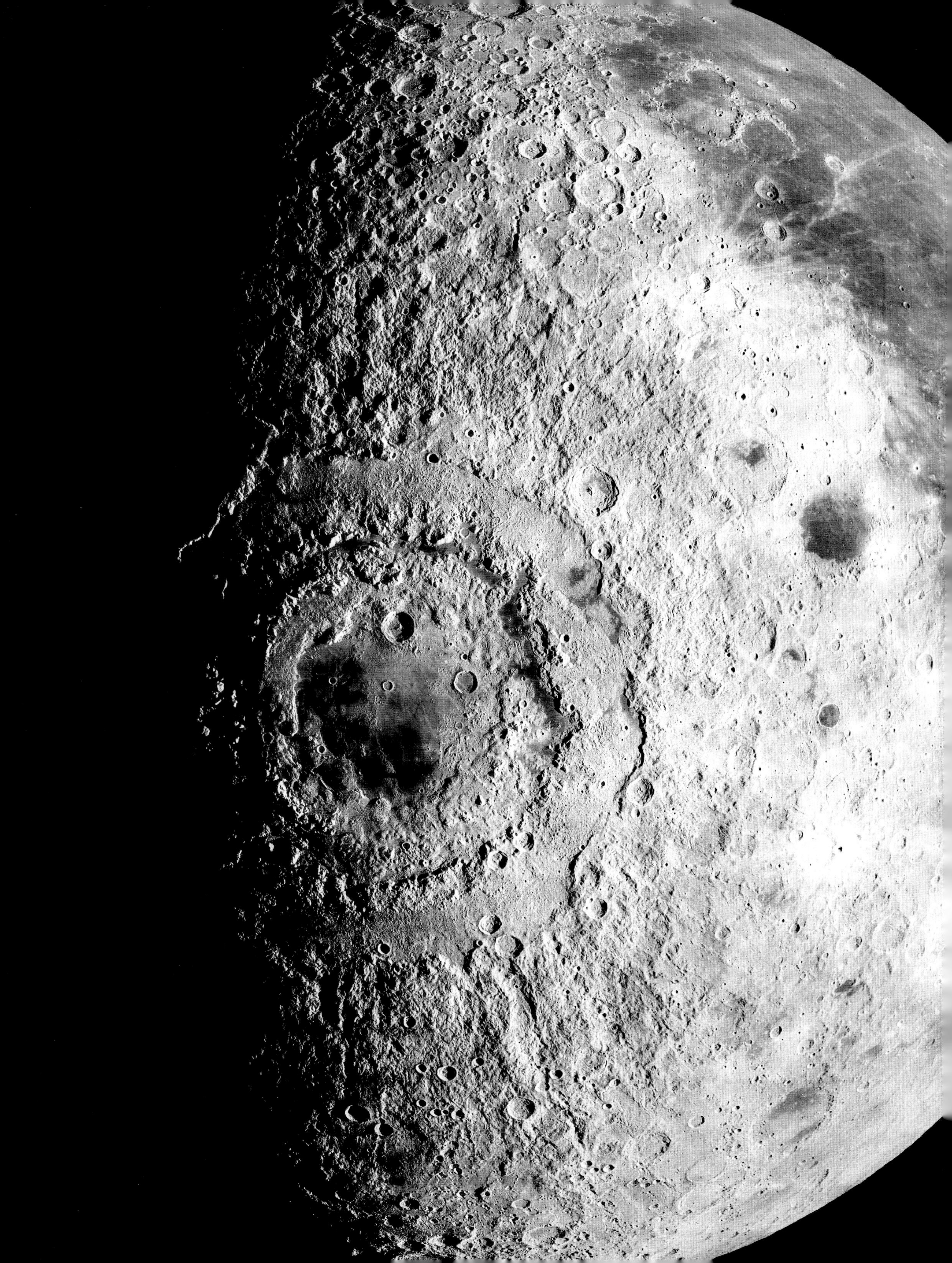

formed remained a cloudy issue until Lunar Orbiter 4 returned the breathtaking image of the 930-kilometer-wide Orientale basin. Being the freshest basin on the Moon, the structure of Orientale has not been modified by subsequent impacts, as have many of the classical near-side lunar basins. This key image linked many previously seen but not fully understood features. Once the big picture of this fresh basin was revealed, it was easy to see similarities in other near-side basins. The issue showed once again the danger of looking too closely at the Moon's geology. Individual features have their interesting points, but an understanding of how the Moon's geology evolved requires a broader view to see how all features relate to each other.

At a combined cost of $905 million (in 1960s dollars), Ranger, Surveyor and Lunar Orbiter were brilliant technical achievements. But after Surveyor 7, an American presence on the Moon continued for the next five years only through the manned landings in the Apollo program.

February 7, 1968, saw the launch of another Soviet Ye-6LS orbiter, but the craft fell back to Earth when the 3rd stage of the SL-6 Molniya booster ran out of fuel 525 seconds into the flight.

The Zond 4 test flight of the Soviet manned circumlunar spacecraft was successfully launched by its Proton rocket on March 2, 1968, but the craft did not go to the Moon. Instead, Zond 4 looped out to a lunar distance on the opposite side of Earth from where the Moon was. Soyuz 7K-L1 No. 6L returned to Earth, but its guidance system failed, and the craft was off course and would not land inside the USSR. Flight controllers commanded the spacecraft to self-destruct at 10 kilometers of altitude over the Atlantic Ocean, 200 kilometers west of equatorial Africa.

The only successfully launched Soviet lunar mission in 1968 was initially a mystery. When Luna 14 departed Earth on April 7, 1968, the announced mission objectives sounded superficial. On April 10, it decelerated into a 160-by-870-kilometer lunar orbit and became the fourth Soviet lunar satellite, circling the Moon every 160 minutes. Luna 14 was tracked by the new TNA-400 antenna complex at Simferopol in the Crimea as it mapped the lunar gravitational anomalies found by previous orbiters. Luna 14 also carried a motor drive for the future Ye-8 lunar rover to test how the device's materials and lubrication characteristics would perform in the space environment.

Two weeks after the Luna 14 launch, Soviet engineers tried to launch another unmanned test of their Zond circumlunar spacecraft. Soyuz 7K-L1 No. 7L rose from Baikonur on April 22, 1968, but failed to reach space when the Proton second stage prematurely shut down and the rocket crashed.

The Soviets Finally Loop the Moon

On September 14, 1968, luck was finally with Soviet lunar efforts when a Proton rocket launched Soyuz 7K-L1 No. 9L. Designated Zond 5, the craft carried two tortoises and smaller life forms on a circumlunar flight that looped 1,950 kilometers above the Moon on Sept 18. Three days later, the craft returned to Earth but missed its landing site inside the USSR and was recovered in the southern

Facing page: One of the most definitive of the lunar science images acquired by Lunar Orbiter 4 was this handsome portrait of the Orientale basin, normally seen edge on from Earth. Lunar Orbiter's overhead perspective confirmed impact origin of the large basins now cradling the dark lunar maria that form the face of the Man in the Moon.

The politically charged race between the U.S. and USSR to send a man to the Moon was won by the United States when Apollo 8 achieved liftoff on December 21, 1968 and circled the Moon 10 times before safely returning to Earth.

Indian Ocean. Zond 5 became the first craft to reach the Moon and return to Earth and the first to carry living specimens to the Moon and back. American officials feared the next Zond flight would carry Soviet cosmonauts around the Moon before the American Apollo 8 could circle the Moon in December of 1968.

NASA's worries about a manned Zond beating Apollo 8 to the Moon proved unfounded. Zond 6 launched on November 10, 1968, carrying more tortoises and biological specimens and looped 2,420 kilometers around the Moon. Although presented to the public as a successful mission, if Zond 6 had been manned, it would have been a disaster. To alleviate high pressure in the craft's hydrogen peroxide tank, the spacecraft was oriented nose first to the Sun. However, solar heating damaged the hatch seal, and the spacecraft depressurized, killing the biological specimens on board. Upon return to Earth, the guidance system worked perfectly, returning the craft to within 16 kilometers of the launch pad at Baikonur. However, a circuitry fault created by the cabin depressurizing in space prematurely ejected the parachute and the craft crashed to the ground. Film cassettes recovered from the wreckage displayed beautiful photos of the Earth and the Moon, making the mission appear successful.

Throughout the nine-year history of the Luna and Zond programs, the series represented an enormous technical undertaking that enjoyed the full support of the Soviet government. While military personnel were involved in launching the Lunas, the lunar exploration program was basically a civilian operation under the guidance of the Soviet Academy of Sciences. The four Soviet lunar orbiters supplemented the two rudimentary Soviet landers to provide a broad general knowledge about the Moon. But these simple missions were mere pathfinders and provided no technical support for the Soviet manned lunar landing program. At this point, the Soviet manned lunar spacecraft had yet to make a fully successful test flight, and the its manned lunar lander was still untested.

Volumes have been written about the technical genius, the scientific goals, and the pure otherworldly romance of the Apollo manned landings on the Moon. The key individuals who planned and managed the Apollo program and the American heroes who dared to ride a pillar of fire and set foot on another world are revered and immortalized in print and cinema. The scope of this book cannot begin to capture the essence and grit of the national obsession to land a man on the Moon before the end of the 1960s. Thus, the manned Apollo landings are addressed in this volume as scientific achievements like their robotic cousins.

The Apollo manned Moon landings were politically conceived in the 1960s to prove that America was second to none in science, engineering, and industrial might. The depth of national pride in achieving the first manned landing on the Moon on July 20, 1969, was only equaled by the Second World War victory over fascist forces bent on destroying democracy.

The scientific, technical, and political victory of the Apollo 11 landing on the Sea of Tranquility was not without a high price. At $20 billion in 1960s dollars, Apollo was the costliest and most complex technical endeavor ever attempted by humankind. And sadly, despite robotic success on the Moon, she showed she was a harsh mistress when American heroes Gus Grissom, Ed White, and Roger Chaffee lost their lives in the tragic Apollo 1 fire in January 1967.

After national soul searching and technical reevaluation, the Apollo program recovered from the Apollo 1 tragedy and just 21 months later the Apollo spacecraft proved itself with the Earth orbit flight of Apollo 7 carrying Wally Schirra, Donn Eisele, and Walt Cunningham. With renewed confidence, NASA then did what a decade before was unthinkable when the Apollo 8 crew of Frank Borman, James Lovell, and Bill Anders orbited the Moon on Christmas Day, 1968.

But in 1968, the race to land a man on the Moon was not a sure bet. NASA gained more confidence with each step-building success with both robotic explorers and tests of the Apollo hardware, but the unknown of what the Soviets were up to, and their two recent unmanned Zond circumlunar flights worried many American officials. Because of the uncertainty, a manned landing was still very much a technical and political race with the Soviets.

A month after Apollo 8's triumphant return to Earth, the Soviets again tried to loop an unmanned Zond around the Moon. Again, the balky Proton rocket prevented Soyuz 7K-L1 No. 13L from reaching orbit when both its second and third stages prematurely shut down.

The landing success of Luna 9 in 1966 was the culmination of a half-decade effort begun by Sergei Korolev. Clues about a more advanced path for Soviet lunar exploration had been sown more than a decade before. Shortly after the first satellite, Sputnik, was launched in 1957, Soviet authorities released a film showing a plan conceived by Yuri Khlebtsevich in 1955 for a remote-controlled television-equipped robot to explore the Moon. The movie appeared on American television and stills from the movie were prominently displayed in *Life Magazine*. The film gave Americans, already stunned by the unexpected appearance of the Soviet satellite, reason to assume the Soviet Union was on the verge of an all-out scientific assault on the Moon. The success of Sputnik showed Soviet science had the potential of accomplishing such a mission.

Early in the Space Age, Soviet engineers began initial plans to fulfill Khlebtsevich's vision. In 1960 a team of engineers working under Mikhail Tikhonravov at Sergei Korolev's OKB-1 design bureau began studying the concept of both wheeled and tank tread-equipped lunar rovers to be launched by a variant of the giant N1 Moon rocket already on Korolev's drawing boards.

In 1963, Korolev transferred all work on the lunar rover's chassis to the Leningrad All-Union Scientific Research Institute No. 100 (VNII-100), a facility primarily responsible for designing tanks for the Soviet Army and headed by Chief Designer Aleksander L. Kemurdzhian. By July 1964, a 900-kilogram rover had been designed to support the Soviet manned lunar landing program. The following May the rover's design was complete, but after a realignment of Soviet spaceflight responsibilities, Korolev transferred automated lunar and planetary spacecraft to the Lavochkin Design Bureau.

When Lavochkin inherited the lunar rover project, it was officially designated the Ye-8. The N1 rocket existed only on Korolev's drawing boards and Babakin and Kemurdzhian decided to redesign the rover from scratch so it could use the existing UR-500K Proton being flight tested by rival rocket builder Vladimir Chelomey. To reach the Moon, the Proton would be topped with the Block D stage, hardware borrowed from Korolev's early manned circumlunar plans.

Rovers Integrated into Soviet Moon Plans

By March 1966, the Ye-8 officially became part of the Soviet manned lunar program. The craft, after scouting for a suitable landing site, would serve as a radar beacon to allow precision landing by a manned spacecraft. Luna 9 data about the apparent firmness of the lunar surface allowed the use of wheels instead of tank treads. By late 1966, the design of the Ye-8 was finished but bureaucratic inefficiencies delayed completion of the paperwork for a year. The spacecraft consisted of a landing stage called the KT stage and the rover was designated the Ye-8L.

In early 1967, the Babakin Design Bureau proposed using the KT landing stage to carry a small sample return spacecraft to the Moon. The Soviet Military Industrial Commission met on October 30, 1968, to decide whether to accelerate the development of the Ye-8-5 sample return mission to fly before the manned American landing reached the Moon sometime in 1969. However, at the time, Lavochkin had not yet built any hardware for the mission.

The Ye-8-5 sample mission was not part of the original Soviet lunar plan and added complexity to the overall Moon effort by draining resources from other projects. Despite this, the political expediency of ensuring a Soviet craft was first to return from the Moon made the development of the sample return mission a priority. Now, two separate third-generation Luna spacecraft were in parallel development, each using the KT landing stage, but with different mission goals. The rover had, at this point, been under design for nearly a

decade and test hardware was being fabricated. At the same time, a new sample return spacecraft was being created at a furious pace to maintain some degree of Soviet political challenge to Apollo. This progressed in the face of the evident failure of the Soviet manned lunar program to perfect the needed hardware to get cosmonauts to the Moon before Apollo landed.

In 1968, working models of the rover were tested in a simulated lunar landscape near Simferopol in Crimea and two teams of rover drivers perfected their control techniques. Test firings of the KT landing stage rocket motor, designated the KTDU-417, took place at the rocket engine test facility in Zagorsk.

The Ye-8L rover, officially called Lunokhod (pronounced "Luna-hod," meaning "Moonwalker" in Russian) consisted of an elliptical 2.1-meter-long bathtub-shaped magnesium alloy instrument compartment that stood 1.35 meters high on a chassis with eight 51-centimeter wire mesh wheels. Titanium blades on the wheels acted as cleats to help propel the vehicle through lunar dust. The Lavochkin Association oversaw the design of Lunokhod, but the Krzhizhanovsky Power Institute in Moscow developed the eight independently

The USSR failed to send men to the Moon, but they did perfect the 756-kilogram, 2.1-meter-long eight-wheeled *Lunokhod 1* rover, which traveled 10.5 kilometers on Mare Imbrium between November 1970 and September 1971.

powered two-speed wheel hubs under the direction of A. I. Moskvitin. The drive system was designed to allow movement at either 0.8 or 2 kilometers per hour even if two of the wheel hubs were disabled on the same side of the vehicle. Normal movement on the Moon was expected to be in short bursts covering between 100 and 200 meters per hour.

The tub-shaped instrument compartment was covered by a convex lid lined with solar cells on the inside. The lid was opened during the day and provided up to a kilowatt of power to charge the rover's batteries and closed at night to retain warmth. During the day, the rover radiated excess heat from the top of the instrument section. Temperature control at night was achieved by a combination of thermo-vacuum insulation and fan-circulated gas warmed by an internal radioisotope heat source powered by 11 kilograms of polonium-210.

Lunokhod sprouted numerous antennas and television cameras, giving it the appearance of a silver insect. Two forward-looking television cameras transmitted a 50-degree wide stereo view every 10 seconds enabling the craft's earthbound operators to monitor and steer the vehicle. A five-man team that included a commander, driver, navigator, systems engineer, and radio operator remotely drove Lunokhod. Control of the rover would be carried out only when the Moon was within sight of the receiving station in the Crimea. The two teams would serve alternating two-hour shifts. Two drivers were chosen, Gabdulkhay Latypov and Vyacheslav Dovgan. They were to be guided in the remote driving duties by mission commanders Nikolai Yeremenko and Igor Fyodorov, who would sit behind them at the control console.

Control of the moving robot was complicated by glare from the Sun, black shadows, lack of depth perception, and a three-second delay in round-trip communication. Lunokhod could be several meters ahead of what the slow-scan television pictures were showing the driver. If the commander sensed there was a potential problem, he could press a button and stop the Lunokhod in its tracks three seconds later.

Four panoramic cameras, used only when the rover was parked, produced 500-by-6,000-pixel images of the local terrain, Sun and Earth for navigation purposes. The most scientifically interesting instrument was the Roentgen Isotopic Fluorescence Method of Analysis (RIFMA). This instrument focused a beam of electrons on a rock sample to induce the emission of X-rays, the spectrum of which would reveal the concentrations of aluminum, silicon, magnesium, potassium, calcium, iron, and titanium.

Lunokhod's zenith-looking X-ray telescope could take exposures of up to six hours while the rover was parked. This is 1,000 times that possible by sounding rockets and resulted in a 10-fold increase in sensitivity over similar satellite instruments in use at the time. At the time of Lunokhod's mission, little was known about celestial X-rays. Data from Lunokhod's X-ray telescope would play an important part in understanding the celestial puzzle of the expansion of the universe.

The rover carried a 3.7-kilogram French laser retroreflector consisting of 14 ten-centimeter tetrahedral corner cube prisms mounted above the forward cameras. Laser shots, high-power burst of laser energy, were fired at the rover from the 262-centimeter Shajn telescope at the Crimean Astrophysical Observatory and the 102-centimeter telescope located at Pic du Midi, France. Precise timing of the laser reflections allowed measurement of the Earth-Moon distance to within centimeters.

Soviet Lunar Efforts Falter

Although spaceflight history shows that a Soviet robotic sample return mission was the first to fly to the Moon, the secret reality of the Soviet Moon effort was that the Ye-8L rover was the first to lift off from Baikonur. The first flight model rover, Ye-8L No. 201, was rushed to a February 19, 1969, launch in an attempt to place a Soviet presence on the Moon prior to the Apollo landings. Fifty-one seconds after liftoff, the shroud covering the rover fell apart due to aerodynamic vibrations and the booster exploded 3 seconds later. Debris from the booster and portions of the rover itself fell 15 kilometers from the launch site. Despite an intensive search of the debris area, the radioactive polonium-210 package was never found. It turns out the radioisotope package had been found by soldiers at the Baikonur Cosmodrome who spirited the assembly away to warm their frigid barracks during the bitter winter of 1968. Eventually the dangerous package was confiscated by authorities.

Two Soviet N1 Moon rockets stand on their launch platforms at the Baikonur Cosmodrome. Although the N1 had a liftoff thrust of 10 million pounds, or 2.5 million pounds more than the Saturn V that propelled Apollo to the Moon, the Soviet rocket was less efficient, and its lunar payload was 20,000 pounds less than the Saturn V. All four N1 test flights failed before the program was canceled in 1974.

the lunar probes. Fortunately, the rocket engineers were able to increase the Proton's lifting capacity.

An ingenious scheme to simplify the Luna sample return spacecraft came from Dmitry Okhotsimsky with the Institute of Applied Mathematics of the Academy of Sciences. He devised a "passive" correction-free flight trajectory only possible on the return from the Moon. Just as the first Luna landers used a simple ballistic trajectory to intercept the Moon at a vertical angle at 63 degrees west longitude on the lunar globe, the new Lunas used a direct three-day ballistic trajectory to return to Earth. To accomplish this, the craft had to liftoff vertically from the Moon at 56 degrees east longitude along the equator and accelerate to 2,700 meters per second. This assured a ballistic trajectory that would automatically intercept Earth's atmosphere. Complicating this simple navigation was the effect of lunar libration which slowly sweeps a circular route around the critical 56-degree east latitude launch location. Luna needed to land on a spot that will have moved, by libration, to the proper liftoff point after securing its soil samples. The study of lunar gravitational anomalies by the Luna 10, 11, 12, and 14 thus became of paramount importance for a precision lunar landing.

The KT landing stage used the brute force of a 265-second engine firing to cancel the lunar orbital velocity and drop to the surface with no horizontal velocity. A radar altimeter controlled the descent engine to slow the craft to 2.5 meters per second at an altitude of 20 meters where the final descent used the stabilizing vernier engines to minimize disturbing the surface.

Upon landing, twin facsimile cameras returned a stereo image of the surface within range of the soil sampling mechanism. An hour after landing, the cylindrical Soil Sample Assembly (SSA) containing the core sample drill was lowered to the ground by a boom pivoting at the base of the lander. The sampling mechanism reached 0.8 meters beyond the lander and could sweep a 100-degree arc. The drill stem consisted of a hollow tube with teeth cut into the leading edge that could reach a depth of 35 centimeters and cut into solid rock. After the drilling was complete, the SSA pivoted up to the lunar return capsule and the drill head rotated 180 degrees. The core sample was then transferred into an ampule and sealed in the return capsule.

Shortly before entering Earth's atmosphere, the spherical sample return capsule was released from the ascent stage. Heavy batteries and transmitting equipment were placed at the bottom of the capsule to offset the center of gravity and stabilize it, placing the thickest portion of the heat shield against atmospheric heating. The Luna slammed into the atmosphere like a bullet, reaching a peak deceleration of 350 Gs.

Once in the lower atmosphere, a bright orange parachute was deployed, flexible recovery beacon antennas uncoiled, and a cloud of metallic needles was released to mark the location on search radars. In case of a water landing, two sausage-shaped balloon floats inflated to keep it upright so the antennas would remain above water. An oscillator circuit broadcast a cricket-like recovery beacon signal.

As the flight of Apollo 11 approached in the summer of 1969, the Ye-8-5 had not yet been tested in space. Additionally, the poor performance of the Proton booster, with seven failures in the past 13 flights and four straight failures up to April 1969, threatened the project. But the Ye-8-5 still had two chances to beat Apollo 11 with lunar launch windows in June and July 1969. On June 14, 1969, Ye-8-5 No. 402 launched from the Baikonur Cosmodrome. But Proton problems continued as the Block D fourth stage failed to ignite and the craft crashed in the Pacific Ocean.

The Soviets suffered another setback on July 3, 1969, when they again attempted a secret test flight of the massive N1 Moon rocket. It was hoped to place the unmanned Soyuz 7K-L1S No. 5 into lunar orbit, then return to Earth. Ten seconds after liftoff, the N1 engines shut down and the rocket crashed with an explosion that was so destructive it would take a year to rebuild the launch pad.

In mid-summer of 1969, as the first Apollo manned lunar landing was imminent, American space officials worried that the Soviets would try a space spectacular that would upstage the Apollo 11 landing. It had been over a year since the last Luna probe had flown. Even test activity with the circumlunar Zond moonship remained strangely quiet. NASA's fear was that an unmanned Luna craft would land just ahead of Apollo, grab a sample of lunar material and return it to Earth before Apollo. Such a feat would be a tremendous Soviet propaganda victory in the Moon race.

A Race with Apollo 11

In fact, this is exactly what the Soviets planned. In July, another lander, Ye-8-5 No. 401 was prepared. This spacecraft had mass problems all the way up to its launch date. The ascent stage, dubbed the RYe-8-5, was allotted a mass of 512 kilograms, but still weighed 513.3 kilograms. In last ditch desperation, Babakin ordered the removal of the 1.28-kilogram backup radio transmitter. This bold move was supplemented by a perfect Proton launch and Luna 15 succeeded in leaving Earth on July 13 just days before Apollo 11. Soviet space officials remained tight lipped about Luna 15. Details of its objective, and ultimate fate, remained a mystery for many years.

Luna 15 arrived at the Moon just as the Apollo 11 astronauts were preparing to leave Earth. As British and American tracking stations watched the mission unfold, it became apparent that Luna 15 was indeed racing with Apollo. However, Soviet silence frustrated efforts to understand what was happening.

After a 102-hour flight from Earth, Luna 15 looped behind the Moon and fired its engine, placing it into an initial 55-by-203-kilometer orbit. By now, Apollo 11 was also en route to the Moon. At the very culmination of the U.S./USSR Space Race, the secrecy surrounding the mission of Luna 15 delighted the news media. The fantastic story of the first manned landing on the Moon was amazing, but the race with the Soviet mystery ship made it all more exciting. Rocket burns on July 19 and 20 trimmed Luna 15's orbit to 16-by-110-kilometers, an indication that something significant was about to happen with Luna 15.

On July 20, 1969, Neil Armstrong and Buzz Aldrin achieved humankind's ultimate exploration goal by landing on the Moon as Michael Collins piloted the Command Module *Columbia* in lunar orbit. After 12 years of cosmic

The political and technical race between the U.S. and USSR to land a man on the Moon ended on July 20, 1969, with the successful landing of Apollo 11 on Mare Tranquillitatis. Neil Armstrong and Buzz Aldrin symbolically planted the American flag on the lunar surface.

competition, American astronauts claimed the political and scientific prize by walking on the Moon as a Soviet mystery ship circled overhead. The drama was played on a worldwide stage.

But today we know the Soviet Moon landing mission had been in trouble from the start. The Luna 15 flight director, Georgy A. Tyulin, was a retired artillery general and First Deputy Minister of General Machine Building. He soon learned the trials of fixing a sick spacecraft in flight. One day after launch, unusually high temperatures in the ascent stage propellant tanks presented the possibility of an in-flight explosion. A quick solution was to turn the ascent stage away from the Sun and keep it in the shadow of the KT stage.

As Apollo 11 left Earth for its appointment with destiny, Soviet mission controllers were in a desperate fix. The restricted zone where they needed to land to successfully navigate back to Earth turned out to be more rugged than anticipated. While NASA's photomaps of the Moon were not a secret, apparently Luna mission planners did not have access to the lunar topography available to Apollo managers. Instead of committing to a landing and beating Apollo 11 to the Moon, mission controllers spent another four days analyzing data while a plan was developed to guide Luna 15 to the surface.

When the Apollo 11 lunar module, *Eagle*, carrying astronauts Neil Armstrong and Buzz Aldrin, touched down on the Sea of Tranquility at 9:59 p.m., the politically charged manned race to the Moon was won hands down by the

Americans. Soviet plans, changing by the moment, called for Luna 15 to begin its descent about two hours after Apollo 11 landed on the Sea of Tranquility. This still allowed the craft to grab a soil sample and return to Earth with Apollo 11 nipping at its heels all the way home. However, delays in mapping the landing site's terrain left uncertainties about the landing zone. Tyulin further delayed the Luna 15 landing by another 18 hours while waiting for his engineers to approve the landing attempt.

Several hours after *Eagle* landed, Neil Armstrong stepped onto the plains of Mare Tranquillitatis and changed the course of human destiny. When Armstrong uttered his famous words, "That's one small step for a man, one giant leap for mankind," he forever established that humankind was no longer bound to the blue and green planet we call Earth.

As spectacular an achievement the first manned landing on another world was, in the scientific scheme, the data gathered by the Apollo landings are merely more lunar datapoints collected during the first 10 years of the political Space Race with the former Soviet Union. Armstrong and Aldrin spent less than three hours on the lunar surface and never explored more than 100 meters from the lunar module. They gathered 21.6 kilograms of samples, including 50 rocks and two core tubes that penetrated 13 centimeters.

On Luna 15's 52nd orbit on July 21, 1969, about two hours before Apollo 11 was to leave the lunar surface, the unmanned Soviet lander initiated its descent to the surface. Contact was lost while the craft was still two minutes from touchdown. A short time later, an official announcement stated that while the Apollo 11 astronauts were still on the Moon, Luna 15 had reached the surface in a preset area. The statement fell short of admitting that Luna 15 had crashed. Analysis by Jodrell Bank, measuring the Doppler shift of Luna's signals, deduced the craft had struck the surface at 480 kilometers per hour at 12 degrees north and 60 degrees east on the southern plains of Mare Crisium, approximately 800 kilometers from Apollo 11 on the Sea of Tranquility.

Apollo succeeded beyond the wildest dreams of its designers, but in the glare of Apollo 11's triumph, the Soviet manned lunar program collapsed. The bureaucratic inflexibility of Soviet economics and rigidly controlled USSR industry could not respond to the manned lunar challenge. Unable to develop a workable rocket to power cosmonauts to the Moon, Soviet lunar exploration continued using only robotic machines built by the Babakin Design Bureau and the Lavochkin Association. Ultimately, the Soviet manned lunar lander resided in a museum and the lunar cosmonauts remained bound to Earth.

Ironically, if the Zond 7 spacecraft, launched on August 7, 1969, had been manned, it would have likely succeeded in sending a Soviet cosmonaut around the Moon within weeks of Apollo 11's landing on the Moon. However, due to continued unreliability of the Proton rocket and the crash of Zond 6, the test of Soyuz 7K-L1 No. 11L only carried four tortoises and other biological specimens. On August 11, Zond 7 looped within 1,984 kilometers of the Moon's surface while performing radiation studies and photographed the Earth and Moon. A week after launch, Zond 7 made a successful landing 50 kilometers south of Kostanay, Kazakhstan.

Rocket Woes for Luna

In September 1969, the cause of the Luna 15 failure had not yet been determined, but Tyulin opted to launch another sample return mission. Ye-8-5 No. 403 lifted off on September 23, 1969, and entered Earth orbit. However, the Block D fourth stage failed to reignite and the craft was trapped in Earth orbit as Cosmos 300. Several orbital maneuvers followed, likely to be engineering tests of the Luna lander's engine.

Ye-8-5 No. 404 was launched a month later on October 22 and again successfully reached Earth orbit. After the Block D stage fired to boost the craft into a translunar trajectory, telemetry became abnormal, then ceased. After two hours, the Kamchatka tracking station revealed the probe had crashed into the ocean near Australia.

While Soviet engineers grappled with the reliability of their Proton rocket, scientists at the Lunar Receiving Laboratory at the Houston Space Center analyzed the prized lunar samples. Within Apollo's rock boxes were lunar basalt samples that contained more titanium than Earth's basalts, breccias — rocks composed of crushed material fused together by the heat and pressure of a meteor or asteroid impact — as well as one highland rock, composed of anorthositic minerals. The basalts were between 3.6 and 3.9 billion years old and the highland rock implied that between 4.3 and 4.5 billion years ago, the Moon was molten, allowing the lightweight anorthosite to float to the top of the Moon's crust and form today's highland areas.

Three major lunar discoveries were deduced from the initial Apollo samples, and quickly banished the pet theories of both Tommy Gold and Harold Urey. The first discovery was confirmation that the lunar surface is hard. Previous unmanned landers had not sunk into the powdery dust postulated by Tommy Gold, and neither did the *Eagle*. About half of the regolith, or meteorite-pulverized lunar soil, was composed of particles smaller than 60 microns and 10 percent of the regolith was particles smaller than 10 microns. Collectively, they easily supported Armstrong and Aldrin, whose boot prints only sank a few centimeters into the lunar soil.

The second discovery was confirmation of the Surveyor lander's detection of basalts, indicating the Moon was once hot, and not a body that accreted cold and never differentiated. The presence of basalts further supported the idea that the Moon was once covered by a global magma ocean. The basalt also confirmed the Moon is a differentiated body where minerals melted and mixed within the Moon, counter to Harold Urey's cold Moon theory.

The third Apollo 11 discovery was that maria are lava flows deposited by volcanic eruptions and not impact melt created by asteroids, as Urey also believed.

Despite being struck by lightning twice in the first minute after launch, the Apollo 12 mission progressed smoothly, and Pete Conrad and Alan Bean landed the lunar module *Intrepid* within 200 meters of the Surveyor 3 spacecraft on Oceanus Procellarum on November 19, 1969. The Apollo missions had now twice fulfilled President Kennedy's challenge to land a man on the Moon by the end of the decade. During two separate moonwalks, the crew collected 34 kilograms of samples, including 45 rocks and core tubes that penetrated

40 centimeters into the surface. Conrad and Bean removed the television camera from the Surveyor 3 lander and returned it to Earth for study.

Apollo 12 landed on a ray structure extending from the Copernicus crater 400 kilometers to the north of the landing site. Analysis of the Copernican ejecta dated the formation of Copernicus at 800 million years ago. All the recovered rock samples were basalt and breccias. The basalt had less titanium than the Apollo 11 samples and were between 3.1 and 3.3 billion years old, demonstrating that lunar volcanism was not a one-time event. Some of

On November 19, 1969, Apollo 12 landed 200 meters from the unmanned Surveyor 3 spacecraft. During their second moonwalk on November 20, astronauts Pete Conrad and Alan Bean examined Surveyor 3 and removed the craft's television camera and returned it to Earth.

Apollo 14 commander Alan Shepard examines a rock outcrop on the Fra Mauro Formation between Mare Imbrium and Mare Cognitum.

Apollo 14 Reinvigorates Manned Exploration

Lunokhod 1 remained in operation while Apollo 14 returned American astronauts to the Moon. After a successful launch on January 31, 1971, Alan Shepard and Edgar Mitchell piloted their lunar module *Antares* to a February 5 touchdown within 30 meters of its target spot on the Fra Mauro Formation north of the Fra Mauro crater. This region was believed to be rich in materials thrown from the Imbrium basin impact. The mission's geologic target was the Cone crater, about a three-kilometer round trip from the landing site. The confusing and rock-strewn terrain around the Cone crater misled Shepard and Mitchell, and although they were within 20 meters of the crater's rim, it went unrecognized and they never reached their desired location.

Although the destination of the Cone crater was not reached by Shepard and Mitchell, the mission was a geologic success. During their two days on the Moon, 42 kilograms of samples were collected. Breccias and impact melt rock that originated in the Imbrium basin were determined to be 2.93 billion years old. The basalts collected were rich in aluminum and potassium and were 4.04 billion years old. More samples of the KREEP geochemical component discovered by Apollo 12 were also recovered.

The next lunar explorations saw Apollo 15 depart for the Moon on July 26, 1971. Apollo 15 was the first mission to use an electric-powered rover to transport the astronauts to distant geologic targets. On July 30, 1971, Dave Scott and James Irwin piloted the lunar module *Falcon* to a touchdown near the northeast bend of Rima Hadley at the base of Montes Apenninus on the eastern Mare Imbrium. The regional basalts were similar to those found by Apollo 12 and were 3.3 billion years old. Apollo 15 recovered what has been nicknamed

Astronaut Dave Scott tests the electric "Moon buggy" rover carried by Apollo 15 that allowed him and Jim Irwin to explore up to five kilometers from the landing site.

so hard the drill began to bind. The resulting sample was only half the size of the Luna 16 sample.

After 27 hours, 39 minutes on the Moon, the ascent stage lifted off on February 23. Two days later, the return capsule parachuted into the height of a nighttime blizzard, where the its beacon was spotted by a recovery aircraft. After the blizzard passed the next morning, a helicopter found the capsule and its orange parachute in deep snow on an island in the Karakengir River, 40 kilometers northwest of Zhezqazghan, Kazakhstan.

Luna 20's 50-gram sample was light gray compared to the dark black slate color with a metallic glitter of the Luna 16 material. When the lunar material was emptied into the receiving tray, its appearance was likened to the ash of a big cigar or a roll of gray cement powder with grains slightly larger than poppy seeds. The deeper end of the core sample contained particles the size of a pea. Using mass spectroscopy, traces of 70 chemical elements were found. The sample shook down to a density of 1.7 to 1.8 grams per cubic centimeter. The samples were thought to be a billion years older than those obtained at the Luna 16 site.

No samples prior to Luna 20 had such a high concentration of aluminum and calcium oxides. Mare samples from Luna 16 only had one to two percent anorthosite, while Luna 20's highland samples contained 50 to 60 percent anorthosite rocks consisting mainly of feldspar. Of the iron containing minerals, 36 percent were olivine, while 57 percent were pyroxene. The new sample was only one percent ilmenite while Luna 16's sample was 10 percent ilmenite.

Over 80 research organizations examined the Luna 20 sample, with some of the analysis done at facilities in the United States, France, Czechoslovakia and Hungary. The Luna 20 results, when compared with Apollo 16 and 17 results, indicated that old crustal rocks in the highlands are primarily anorthositic gabbros and related type rocks while the mare areas sampled by Luna 16 and Apollo 11 contain mainly basaltic rocks. Luna 20 also found a larger quantity of metallic iron than did Luna 16. This iron was unusual in that it did not rust in the air. Laboratory experiments showed this type of iron is sublimated out of basalt rocks at high temperature in a vacuum.

Although the lunar highlands cover five times the area of the maria, Apollo 16 was the first to land in a highland area. On April 21, 1972, John Young and Charlie Duke piloted the lunar module *Orion* to a touchdown on what was thought to be a volcanic region west of Theophilus crater. The mission objective was to sample both a Cayley plain, so named after similar flat plains near the crater Cayley, and the Descartes formation. Nearby Stone Mountain was also thought to be a volcano. However, when John Young looked out *Orion*'s window after touchdown, he did not see a smooth volcanic plain, he saw many rocks indicating the region was filled with ejecta from nearby impacts.

During their three-day stay on the Moon, Young and Duke used a rover to travel 26.7 kilometers and pick up 95.7 kilograms of samples from up to 4.6 kilometers from the landing site. The pair traveled to the South Ray crater and sampled the two-kilometer wide, 680-meter-deep formation. They also visited the North Ray crater, where they recovered a sample that originated

230 kilometers below the surface. They also sampled a 20-meter-long boulder named "House Rock" that was believed to have been excavated from 200 meters below the surface by the impact that created the North Ray crater.

Apollo 16 astronaut Charlie Duke investigates House Rock, believed to have been excavated from 200 meters below the surface by the impact that created the North Ray crater.

In all, Young and Duke recovered 731 rock and soil samples and a 2.2-meter-deep core sample. Almost all samples were breccias with some highland rocks, one of which was 4.36 billion years old, and not the volcanic basalt anticipated before the mission. The small amounts of basalt found was thought to be ejecta from Mare Nectaris.

Before leaving lunar orbit, PFS-2, another Particles and Fields Subsatellite like that launched from the Apollo 15 service module, was released by Apollo 16. However, due to difficulties with the service module engine, Apollo 16 departed the Moon early and was not able to raise its orbit before releasing PFS-2. Because of the lower orbit, the recently discovered mascons altered the orbit of PFS-2 and it crashed into the Moon after only 424 orbits.

The final blow to the Soviet manned lunar program came on November 23, 1972. The fourth test flight of the Soviet N1 Moon rocket was to loft an unmanned Soyuz 7K-LOK into lunar orbit, then return to Earth. Ninety seconds after liftoff, the central six engines in the N1's cluster of 30 first stage engines shut down as scheduled. However, the hydraulic shock ruptured a fuel line, creating a fire in the rocket's tail and engine 4 exploded. The vehicle broke up 107 seconds after launch and crashed. Before a fifth test launch could be conducted, the N1 program was canceled in May 1974.

The final Apollo landing targeted Apollo 17's lunar lander *Challenger* to the Taurus–Littrow Valley on the southeastern shore of Mare Serenitatis. Astronauts

Apollo 16 commander John Young salutes the American flag while taking advantage of the weak lunar gravity to jump almost a meter above the surface.

Gene Cernan and Harrison Schmitt, the first PhD geologist to go to the Moon, landed on December 11, 1972, and again used a rover to explore the region for three days. The pair traveled 35.7 kilometers reaching up to 7.6 kilometers from the landing site. Along the way, Cernan and Schmitt collected 110.6 kilograms of samples and left a seismic experiment that detonated eight explosive packages after they departed the Moon.

Apollo 17 samples dated the Imbrium basin impact between 3.83 and 3.91 billion years ago and the Serenitatis basin impact between 3.93 and 4.22 billion years ago. Several of the Apollo 17 rock samples were dated as 4.2 to 4.5 billion years old and originated in the lower half of the Moon's crust when the early magma ocean was solidifying, then were excavated by the basin impacts.

One of the mission's exciting finds was orange soil near the Shorty crater. The orange soil was 3.6-billion-year-old volcanic glass that originated from melted material 400 kilometers below the surface before being ejected as a fire fountain. Initially, it was thought Shorty was a volcanic vent, but it turned out to be an impact crater younger than the volcanic glass around it. The rocks in the Taurus–Littrow Valley were titanium-rich mare basalts that originated from 130 to 220 kilometers below the surface before erupting and solidifying 3.7 to 3.8 billion years ago. Seismic data from the explosive packages left by Apollo 17 showed the basalt layer under the landing area was between one and 1.4 kilometers deep.

Without the specter of Soviet competition driving the Apollo program, Congress and the White House canceled later science-oriented Apollo landings. Hardware built for Apollos 18, 19 and 20 became museum pieces to be gawked at by tourists and mourned over by those who dreamed of reaping American scientific glory. After Apollo 17 lifted off from the Moon in 1972, America abandoned further lunar exploration for the next 26 years. For the rest of the 1970s, the only vehicles approaching the Moon carried a red star.

The only geologist to fly to the Moon, Apollo 17's Harrison Schmitt, explores the Taurus–Littrow Valley on the southeastern shore of Mare Serenitatis.

Solitary Soviet Lunar Exploration

Now having a monopoly on Moon exploration, Soviet propagandists stressed that manned expeditions to the Moon were too expensive and future robotic Lunas would explore our natural satellite cheaply and without risking lives. Despite the Soviet rhetoric, few missions remained in the Luna series. At this point the political competition with Apollo was over and the Soviet manned lunar program had tapered into oblivion. The USSR had lost the political race to the Moon and the victorious Americans now also abandoned our natural satellite. The Luna program was proceeding at a reduced pace after the 1971 death of Georgy Babakin. Soviet bluster aside, the Luna program was also ramping down despite its ability to finally produce real science.

Luna 21, carrying the *Lunokhod* 2 rover, was launched on January 8, 1973. The lunar landing was almost aborted when engineers received indications that one of Luna 21's landing legs had failed to deploy. Issues with the rover's power supply also required the solar panel to be opened in flight to recharge its batteries. Prior to launch, a short circuit caused an avionics box to fail, requiring the removal of many of *Lunokhod* 2's components. Subsequent tests of the electrical systems shortened their expected lifetime and foretold future issues on the Moon.

Lunokhod 2 was similar in appearance and operation to its predecessor but had grown to 836 kilograms. An extra-high-resolution navigation television camera mounted at eye level returned an image every three seconds, giving the driver on Earth more confidence. A magnetometer mounted on a 1.5-meter boom in front of the vehicle, nicknamed "The Cobra," could measure the magnetic properties of individual rocks. Another new device was an astrophotometer that measured the brightness of the sky over the Moon during the day and studied zodiacal light and light levels of the Milky Way at night. The new rover also carried the improved RIFMA-M X-ray spectrometer to analyze soil chemistry and the Rubin 1 photo receiver for detecting laser bursts from Earth.

Lunokhod 2 achieved lunar orbit on January 12, 1973, and descent to the lunar surface began on January 16 with touchdown inside the crater Le Monnier on the eastern shore of Mare Serenitatis. The landing was only 180 kilometers from the Apollo 17 landing site, where the last humans on the Moon had departed a month before.

The first television images from the surface showed the lander had come to rest on a level plain between two small craters. The low western foothills of the Taurus Mountains were visible six kilometers away. The same two teams of controllers that had guided *Lunokhod 1* were again in control. Using the experience gained driving that last rover, controllers quickly avoided a 15-meter crater and maneuvered to a spot 30 meters away, where they turned the vehicle to face the Sun and charged its batteries for two days. Despite driving experience gained with *Lunokhod 1*, on the first lunar day there was a near collision between *Lunokhod* 2 and the Luna 21 landing stage. An unresponsive control system also almost caused the rover to roll into a crater. After ignoring two stop commands, the vehicle finally responded and made an emergency stop.

Lunokhod 2 traveled twice as fast as its predecessor and was commanded to begin travels to the south. Geologist Alexander Basilevsky again directed the traveling robot's geologic investigations and was allowed more freedom to select scientific targets. Chemical tests by the RIFMA-M spectrometer showed the local lunar soil was similar in composition to that at Mare Imbrium, but generally there were more boulders and rocks. During the first lunar night, the temperature on the Cobra boom fell to -183 degrees Celsius.

The second day's lunar travels moved steadily toward the Taurus Mountains, taking magnetic measurements along the way. The Moon itself has no magnetic field, but lunar mountains and maria have local magnetic fields. At one point the rover climbed an 18-degree slope and entered a 100-meter-wide crater, only to sink to its axles in dust. As the rover zigzagged up the slopes, wheel slippage often approached 80 percent as the wheels churned in the powdery lunar soil produced by meteorite impacts that created the crater and its sloped walls.

Travels on lunar day three and four worked along a 300- to 400-meter-wide, 40- to 80-meter-deep chasm nicknamed Direct Furrow which allowed the study of a cross section of the subsurface layers. Traveling along the western edge, then transversely along the eastern side, Lunokhod established that it was on solid basalt rock that had been split by tectonic activity. As the edge of the fracture was approached, the depth of surface soil declined until there was just a stone rim. Navigation was difficult because of meter-sized boulders along the edge.

The X-ray spectrometer revealed that the lunar soil near the landing site contained 24 percent silicon, 9 percent aluminum, 8 percent calcium and 6 percent iron. As *Lunokhod 2* traveled south across the Le Monnier crater the amount of elemental iron content of the soil declined. Eight kilometers from the landing site, it had dropped to 5 percent while the aluminum content rose to 11 percent. A month later while further away, the iron reading was only 4 percent, the lowest found during *Lunokhod 2*'s travels.

Other science returned was the detection of variations in the lunar magnetic field near Direct Furrow and the existence of residual magnetism in lunar rocks which implied a strong lunar magnetic field in the past. The instrument also measured changes in the minute lunar magnetic field while the rover was stationary. This provided information about the Moon's interior and suggested the Moon possesses a weak magnetic field, a discovery in agreement with Apollo findings. Laser shots reflected from *Lunokhod 2* established the distance from the Earth to the Moon to within 40 centimeters. A shift in the Earth's poles of only 10 centimeters was also detected. Laser shots from the 2.7-meter Harlan Smith telescope at McDonald Observatory in Texas were also reflected by the rover's French laser reflector.

The fifth lunar day began with travels north along the eastern rim of Direct Furrow when contact was abruptly lost for reasons not revealed at the time. Decades later, two versions of the problem were reported. One relates that *Lunokhod 2* overheated and died on the fifth lunar day of operations because of an incident on May 9, 1973, when it was accidentally driven into a small crater. When *Lunokhod 2* tried to back up, the solar cell-lined lid trailing behind the craft bumped the crater wall and a cascade of lunar soil splashed

The unmanned Soviet *Lunokhod 2* rover crisscrossed its own wheel tracks as it zigzagged across Le Monnier crater in 1973. *Lunokhod 2* was the last craft to explore the surface of the Moon in the 20th century.

across the solar cells. As lunar night approached, the rover's lid was closed to keep the craft from freezing. The soil fell from the solar cells onto the radiator on top of the rover's body. The next morning the rover's lid was opened, but with soil covering the radiator, the rover's internal temperature began to rise. When it reached 65 degrees Celsius, the rover's electronics died. The second version relates to continued issues with the electronics that were replaced prior to launch. During the 503rd communications session with the rover, it was noted that the electronics had heated to 47 degrees Celsius. Equipment was shut down to allow it to cool, but contact could never be reestablished.

Lunokhod 2 lived past its 90-day design life, traveled 37 kilometers and performed 50 percent more soil tests than *Lunokhod 1* while producing 86 panoramas and 80,000 television pictures. It additionally made 4,000 laser contacts and performed 740 mechanical soil tests.

The story of *Lunokhod 2* would have ended here except 20 years later Lunokhod and its Luna 21 lander were sold by the Lavochkin Association for $68,500 at a Sotheby's auction in New York to computer gaming entrepreneur Richard Garriott. *Lunokhod 2* and its landing stage are permanently resting on the Moon. Therefore Garriott, who flew to the International Space Station as a paying tourist in October 2008, owned *Lunokhod 2* for the bragging rights. Then, 37 years after contact was lost, NASA's Lunar Reconnaissance Orbiter (LRO) succeeded in photographing the rover's tracks in sufficient enough detail to allow the wanderer to be located on the Moon. *Lunokhod 2* may be mute on the Moon, but now that its location is known, it is still useful for the French-built laser retroreflector. A powerful laser fired from Apache Point Observatory in New Mexico was routinely bounced back by *Lunokhod 2* for experiments in geodesy and celestial mechanics.

Luna Program Winds Down

The sixth Soviet lunar satellite, Luna 22, rose up from Baikonur on May 29, 1974. Three days later it entered a circular 220-kilometer lunar orbit and began a one-year research program expanding on Luna 19's photography, radar altimetry, magnetic field and gamma ray spectrographic studies of lunar composition. On June 9, its orbit was lowered to 25-by-224 kilometers for panoramic photography. Four days later, Luna 22 was returned to a 181-by-299-kilometer orbit to study the perturbations by lunar gravity. After 4,000 revolutions in 14 months, Luna's orbit was again lowered on August 24 to 30 kilometers for more photography, then raised back to a 100-by-1,286-kilometer orbit and the mission continued until November 1975. The long duration of the Luna 22 mission exhibited a new maturity in Soviet space operations.

Luna 23 was the second Soviet Moon flight in 1974 and was intended to be another surface sample return mission. Launched on October 28, 1974, the planned landing site was the same area Luna 15 had aimed for five years before. On November 2, Luna 23 entered a 94-by-104-kilometer lunar orbit. Three days later, this orbit was lowered to 17-by-105-kilometers preparatory to landing.

Luna 23 carried a new LB-09 automatic soil-sampling device designed to sample soil up to two meters below the lunar surface without disturbing the stratification of the layers. Instead of being mounted on a pivoting arm, the improved drill was rigidly mounted to the side of the descent stage and was capable of penetrating deeper into the surface. The new sampler used a percussion drill to penetrate soft material, then switched to rotary drilling to bore through any obstacle. The hollow saw-tooth drill stem had a flexible inner liner which, during the process of drilling into the surface, was stuffed with soil and rock chips like a sausage casing. When the drilling was complete, the 8-by-1,600-millimeter drill liner containing the core sample was coiled like a rope on a winch. The drill mechanism was then hoisted up its mounting framework and the sample transferred and sealed into the return capsule. The drill was swung away to clear the ascent stage for liftoff.

On November 6, Luna 23 touched down on rough territory on Mare Crisium. In an unlucky break, the craft's drilling mechanism was damaged, possibly by a chance impact against a rock while landing, and was unable to complete its task. A reduced research program was conducted for the next three days and on November 9 the mission was abandoned.

A replacement mission to complete the task that eluded Luna 23 was launched on October 16, 1975. However, the Proton booster failed and the new spacecraft never reached space.

Luna 24, the Soviet Union's last flight to the Moon, was launched on August 9, 1976. Identical to the ill-fated Luna 23, the craft entered lunar orbit on August 14. Descent to the surface began after the 53rd orbit on August 18 with touchdown four minutes later. The landing was in southeast Mare Crisium near where both Lunas 15 and 23 tried to land. Shortly after landing, drilling began. A total of twelve cake-like samples were collected, each about 10 centimeters long and 2.5 centimeters thick, from depths reaching two meters.

After a 23-hour stay on the Moon, the return capsule lifted off on August 19, 1976. The Luna 24 descent stage continued to transmit data and test the lunar drill. Three days later, the capsule parachuted into a forest 200 kilometers southeast of Surgut in central Russia.

Early reports from the Vernadsky Institute indicated that the drilling mechanism encountered several distinct layers of deposits with differing densities, possibly due to the ejecta blankets from meteoric impacts dating back three billion years. The automated drill collected both fine dust and small particles. Yuri Zolotov, deputy director of the Vernadsky Institute, said the color of the soil was silver or gray with a brownish tint like the Luna 16 samples, but the Luna 24 soil was lighter and had larger grains. A total of 150 grams were recovered and the sample was broken up into 30-centimeter segments for analysis which revealed 60 different chemical elements.

Three decades after Luna 24 became the last vehicle to soft land on the Moon during the 20th century, it was spotted from lunar orbit by NASA's Lunar Reconnaissance Orbiter (LRO) spacecraft. High-resolution images of the craft's landing site showed Luna 24 came perilously close to disaster by having landed on the rim of an approximately 100-meter-wide crater. Within the same image, Luna 23 rested only 2,700 meters from Luna 24.

Luna 24 marked the end of the initial phase of lunar exploration for both America and the USSR. Just as the American Apollo manned lunar missions were canceled by the government when they achieved an operational maturity, the Soviet Luna missions were also stopped just as they began to return significant scientific data.

Lunokhod 3 was planned but never executed. The Luna program finally ran out of political steam and no resources were allocated to send the already-built rover to the Moon. Rockets were instead applied to Soviet Venus exploration and efforts to reach Mars in competition with the American Viking Mars program. Today, *Lunokhod 3* sits quietly in the Lavochkin museum. Indeed, as late as 1978, a Luna KT landing stage and sample return stage were still in place at the spacecraft processing facility at the Baikonur launch center. In time, these too joined the ever-growing roster of museum pieces that reflected the dashed hopes of lunar scientists.

Curiously, at the end of the golden era of lunar exploration, only three spaceships have ever returned from the surface of the Moon, and none of them were Apollo. The American Apollo Command Modules that carried our astronauts to and from the Moon were merely taxis to get them into lunar orbit. The landing and subsequent ascent from the Moon was aboard the Apollo Lunar Module, which was then abandoned at the Moon. Only the three successful Luna sample return missions alighted on the Moon then returned to Earth.

Looking back at the brief period when lunar exploration was an international priority, we can reflect that so much was accomplished in so little time. With Cold War politics pushing both nations, the Soviet Union and America reached the Moon less than a decade after the Space Age began. Then, after a hard, expensive struggle by thousands of dedicated scientists and engineers from both nations to pave the road to the Moon in the 1960s, the lunar goal was completely abandoned in the 1970s.

Each country had spent billions of dollars in the race to the Moon. While science was the stated goal, what the vast expenditures really purchased was national prestige at a time when propaganda and national image outweighed everything else. When the Moon challenge was no longer a political race, both participants canceled the program and ignored the science that was the originally stated goal. If there is any such thing as a scientific crime, the abandonment of the Moon after two decades of national sweat, heartbreak and enormous expense certainly qualifies as one.

5 The Moon Remains a Harsh Mistress

The liftoff of Luna 24 from the Moon in the summer of 1976 marked the final shot in the classic U.S./USSR Moon race of the 1960s and early 1970s. America had already abandoned the Moon at this point and when Luna 24's return capsule touched down on Russian soil, the USSR also gave up their lunar effort.

Without the excitement of the breakneck political lunar race between the two world superpowers, both public and scientific interest in the Moon waned. The entire 1980s would pass without a lunar exploration mission flown by any nation. During that time what had been learned during the furious dash leading up to Apollo and its subsequent manned landings was analyzed and sorted out. An enormous amount had been leaned about our Moon, but it was soon apparent there were still gaps in our understanding of the Moon's geology, and indeed, understanding the very process by which the Moon was created.

It was also obvious that the expensive breakneck pace of the Space Race with the Soviets was no longer the way to do lunar science. A cheaper, more methodical way to continue exploring the Moon was needed. At the same time, the space programs of other nations besides the U.S. and USSR reached a maturity allowing European and Asian countries to mount deep space missions.

Thus, as the 20th century ended, the exploration of the Moon shifted from the politically driven two-horse race between the Americans and the Soviet Union into a more affordable and efficient effort by many different nations of Earth. As the 20th century ended, Japan joined the roster of lunar faring nations. As the 21st century dawned, the flags of nations that were regarded as third-world countries when Apollo 11 landed were also preparing to fly to the Moon.

Fourteen years after Luna 24 returned to Earth, a craft was launched that would take a curious circuitous route to the Moon and allow Japan to join the exclusive club of lunar explorers. Three years after it was launched, Hiten would impact the Moon, but it was not actually a lunar probe. Before finally reaching the Moon, Hiten would wander space like a modern-day Marco Polo while testing technologies for use on future Japanese lunar and planetary missions.

Initially, the spacecraft was designated MUSES-A. Once in space, it was renamed Hiten after a flying music-playing Buddhist angel. Hiten carried a

hitchhiking subsatellite called Hagoromo, named for the veil worn by the angel Hiten. The subsatellite was destined to enter lunar orbit and was the first of a new class of spacecraft that would use onboard thrusters to slowly maneuver themselves from Earth orbit to the Moon by lazily looping around the Earth in ever-higher orbits that finally intercepted the Moon.

The solar cell-covered drum-shaped 1.4-meter wide, 0.8-meter tall, 185-kilogram Hiten and its 12-kilogram Hagoromo hitchhiker were launched from the Kagoshima Space Center in Japan on January 24, 1990. The Japanese-developed solid-fueled M-3SII placed the spacecraft into a highly eccentric orbit that would allow repeated flybys of the Moon. On March 18, Hiten performed its first lunar flyby and released Hagoromo. Unfortunately, the hitchhiker's transmitter had failed the previous month. Ignition of the subsatellite's orbit insertion motor was observed from Earth and Japan became the third nation to orbit the Moon.

Hiten completed nine flybys of the Moon before being gravitationally captured at the L4 and L5 Lagrange points, where it searched for interplanetary dust trapped at the libration points. No increase in dust was detected. Hiten was then commanded to enter a low lunar orbit on February 15, 1993. Two

Launched on January 24, 1990, Hiten, known in English as the Celestial Maiden, was Japan's first lunar probe and coincidentally the 100th spacecraft launched toward the Moon. After a widely looping three-year trajectory through near-Earth space, the craft impacted the Moon in 1993 between the craters Stevinus and Furnerius.

months later, Hiten was deliberately crashed, making it the first non-U.S. or Soviet craft to hit the Moon. Curiously, Hiten was the 100th Moon mission flown since Pioneer 0 in August 1958.

The Galileo Jupiter probe, launched by the Space Shuttle *Atlantis* on October 18, 1989, also made some interesting discoveries on the Moon. The craft flew past Earth twice to be gravitationally lofted toward Jupiter. During the 1992 flyby, the craft took 53 images of the Moon's far side revealing the previously unrecognized South Pole–Aitken basin, now regarded as one of the largest impact basins in the solar system. Galileo, for the first time, imaged the Earth and Moon together in one family portrait. The image emphasized how dark the Moon is compared to Earth.

NASA's Galileo Jupiter probe, launched in 1989, took a looping trajectory through the inner solar system on its way to Jupiter. On December 16, 1992, Galileo looked back toward Earth and snaped the first photo of the Earth and Moon together in space. The image demonstrated the difference between the dim Moon's seven percent reflectivity and the bright Earth's 30 percent reflectivity.

Clementine Returns America to the Moon

The two-decade hiatus in American lunar flight was broken on January 25, 1994, when *Clementine*, the first lunar probe launched from Vandenberg Air Force Base, California, lifted off from Space Launch Complex 4 aboard a decommissioned Titan II ballistic missile. *Clementine* originated as a military Strategic Defense Initiative Organization (SDIO) program to test the new concept of electronic imaging sensors and morphed into a science program when the original SDIO program was discontinued.

Clementine, a box-shaped 1.88-meter high, 1.14-meter wide, 230-kilogram spacecraft was a hybrid of sorts, using military hardware to do NASA-like science. *Clementine* was to perform a two-month multispectral photographic survey of the entire Moon, perform laser altimetry studies and further investigate the Moon's gravitational anomalies. Upon completion of its lunar mission, *Clementine* would then use its onboard rocket engine to depart lunar orbit and travel to the asteroid Geographos, where it was to perform the first geological survey of an asteroid.

Initially saddled with the dull military acronym of DSPSE, for Deep Space Program Science Experiment, the craft was mercifully renamed *Clementine* after the prospector's daughter in the folk song of the same name. This brightened the public perception of the new lunar mission. *Clementine* was designed, built and controlled by a team of 55 people in only 22 months and at a cost of only $80 million, a pace not matched since NASA's "can-do" era in the 1960s.

Clementine's electronic cameras viewed the Moon in 11 different wavelengths between the ultraviolet and the infrared, allowing analysis of the Moon's geologic makeup by recording how sunlight reflects off the various minerals on the lunar surface. The Lunar Orbiter satellites flown a quarter century before were capable of higher resolution, but their image scale was not consistent over the entire Moon. *Clementine* would now map 100 percent of the Moon at the same image scale. *Clementine*'s high-resolution camera used a 131mm aperture and a 1250mm focal length telescope that achieved between a seven- and 20-meter resolution at the surface. The camera would image in 10-frame bursts and returned long narrow strips of high-resolution imagery that were mosaicked together to map wide areas. A laser altimeter to measure surface

After a quarter century of ready alert, a decommissioned Air Force Titan II ICBM was repurposed as a space launcher and lofted the *Clementine* probe toward the Moon on January 25, 1994.

elevation to an accuracy of 40 meters fired one pulse per second through the same optics used by the high-resolution camera.

Clementine's booster, a converted Titan II ICBM, had spent 25 years with an H-bomb in its nose while in a missile silo in Arkansas. The former military missile thrust *Clementine* into an initial low-Earth orbit on January 25, 1994. After several days in low-Earth orbit, it used its own thruster to boost itself halfway to the Moon. *Clementine* looked back at our planet and relayed detailed photographs of the nightside of Earth showing sparkling city lights. *Clementine* looped past Earth twice to gravitationally slingshot itself toward the Moon and entered lunar orbit on February 20, 1994, and soon settled into its two-month lunar mapping routine.

Over its 71 days in lunar orbit, *Clementine* tirelessly churned out almost a million ultraviolet and visible light images, 620,000 high-resolution images, and 320,000 mid-infrared images. The craft also provided the best data yet on the global composition of the Moon and the composition of mare basalts on the far side.

The laser altimeter provided interesting new data about the recently discovered South Pole–Aitken basin. The giant basin spans half the width of the Moon's far side and had eluded detection by Lunar Orbiter because its huge

The *Clementine* spacecraft started as a Defense Department technology demonstrator for the Ballistic Missile Defense organization and evolved into a NASA Moon probe.

size masked its presence. The *Clementine* laser showed the massive basin has isostatically rebounded to a six-kilometer depth.

Mission control for *Clementine* contained the same high-tech gadgetry as traditional spacecraft control centers, but it was not located in a famous space center. The facility was housed in an old brick warehouse in Alexandria, Virginia, across the street from an undistinguished row housing project. The facility, located in a gritty urban location, was thus dubbed the "Batcave" after the first of the popular Batman movies.

The iconic image of the *Clementine* mission is not one of the Moon or the spacecraft, but of mission scientist Eugene Shoemaker sitting entranced on the floor atop a two-by-two-meter mosaic of the lunar south pole. Shoemaker, lost in image details so subtle the untrained eye could not see them, had spanned the entire gamut of lunar exploration. He started his lunar journey with his 1950s studies of Meteor Crater in Arizona and resolved to be the first scientist and astronaut on the Moon. When he was disqualified from spaceflight by Addison's disease, he helped train the Apollo astronauts in the 1960s. Now he embraced the Moon again 22 years later as a *Clementine* scientist.

After concluding one of the most scientifically productive lunar missions to date, *Clementine* was commanded to leave the Moon on its 349th lunar orbit for its departure to the asteroid Geographos. Although an unqualified lunar success, the craft was an asteroid failure as a computer malfunction accidentally depleted the craft's attitude control fuel. *Clementine* was trapped in Earth orbit in an unstoppable 80 revolution-per-minute spin. By June 1994 the craft's radio signal faded and *Clementine* was indeed lost and gone forever, just like its namesake folk song predicted.

Big Science on a Low Budget

Lunar Prospector was conceived as a low-cost mission that would fly instruments to the Moon to gather non-image data to supplement the photographic archive returned by *Clementine*. Since *Clementine* had been a Department of Defense program, *Lunar Prospector* marked NASA's return to the Moon 26 years after Apollo ended.

The *Lunar Prospector* mission, spacecraft, booster and data collection was carried out for a mere $63 million. The craft was a drum-shaped epoxy-graphite structure standing 1.28 meters tall, 1.37 meters wide and having a fully-fueled mass of 295 kilograms. The circular exterior of the probe was covered with solar cells. Three radially deployed 2.5-meter instrument booms were spaced equidistant around the lower perimeter. The craft was spin stabilized but had six thrusters for course corrections. *Lunar Prospector* had no onboard computer. Each of the experiments was directly controlled from Earth and data

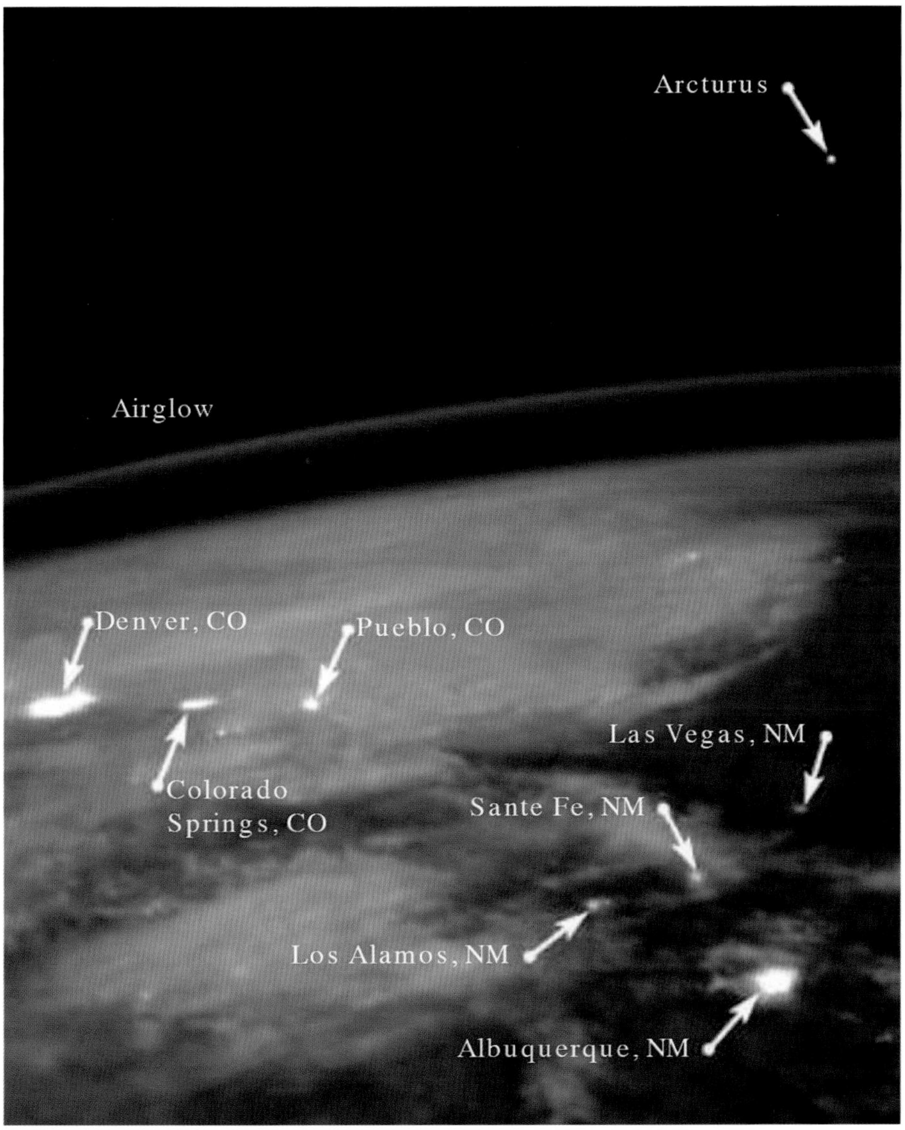

On its way to the Moon, *Clementine* looked back at Earth and used its star tracker camera to image city lights on the nightside of the Earth.

Lunar Prospector was a low-cost mission that orbited the Moon for 19 months in 1998 and 1999 and charted the surface mineral composition and abundance of hydrogen in one of the early searches for traces of water on the Moon.

was downloaded continuously. The craft's limited memory stored data when the craft was behind the Moon.

Lunar Prospector was launched on January 6, 1998, by a solid-fueled Athena II rocket, a low-cost alternative to expensive and complex liquid-fueled space boosters. Previous Moon rockets looked alive as they groaned and seethed in cryogenic vapors on the launch pad. *Lunar Prospector*'s solid-fueled Athena II stood like an inert tree stump. The last American lunar explorer to leave Florida was Apollo 17. In the only night launch in the Apollo program, the Saturn V turned the Florida night into an eerie orange daylight glow as it ascended from Cape Canaveral. Now 26 years later, the Athena II looked more like a fireworks skyrocket than a spacecraft as it barreled into the dark Atlantic sky.

After a 105-hour flight to the Moon, *Lunar Prospector* was maneuvered into a two-hour 100-kilometer circular polar orbit. On January 15, 1998, the craft began its one-year mission studying lunar magnetic and gravitational fields, surveying the surface composition, and searching for evidence of ice at the Moon's poles. In January 1999 the mission was extended while the craft was maneuvered first down to a 30-kilometer orbit, then down to a 10-kilometer orbit to survey the surface at even higher resolutions. Overall, the quality of the surface composition data retrieved by *Lunar Prospector* was 10 times better than had been hoped for.

Lunar Prospector's neutron spectrometer readings showed a consistent dip in the energy of reflected cosmic ray neutrons as the craft flew over the poles, readings consistent with the presence of ice. Analysis of the data gathered from a lower orbit indicated that as much as three billion tons of ice crystals might be in the upper 50 centimeters of the shaded polar regolith. An early deduction from the craft's gamma ray spectrometer was that the distribution of thorium on the Moon is primarily concentrated in the Mare Imbrium area.

High-resolution gravity mapping revealed the existence of three mascons previously unrecognized from the Apollo-era and *Clementine* data. The new-found mascons lie under Mare Humboldtianum, the Mendel-Rydberg Basin, and the Schiller-Zucchius Basin. Mascons are usually associated with mare regions and the subsurface area near Schiller is a cryptomare, that is a mare area that was subsequently buried by ejecta from another later basin. Gravity measurements refined size estimates of the Moon's core, suggesting an iron core with a radius of 300 kilometers.

Before *Lunar Prospector*'s investigations, it was thought the Moon's intrinsic magnetic field was too weak to repel charged particles from the solar wind. It turns out the Moon possesses enough magnetized rocks in its upper layers to create localized dipole magnetic fields, each about 100 kilometers in size. The spacecraft's magnetometer and electron reflectometer found evidence that localized magnetic fields on the Moon can, in unison, create the smallest magnetosphere in the solar system and repel the solar wind from the lunar surface.

The localized magnetic fields are often found at the antipode, or diametrically opposite, of the global location of the younger large impact basins. These magnetic fields are thought to be created by two factors. The seismic shocks from a huge impact passed through the Moon and refocussed on the surface opposite the impact area with enough force to fluidize the surface material, which then resolidified. At the same time, impact ejecta from the basin impact rained down all over the Moon, with some of it traveling all the way around the Moon and striking at the antipode from all directions. The combined shocks induced a magnetic field in microscopic metallic iron particles in the soil. The second factor was ionized gases from impact vaporized rock. The ionized gas expanded all the way around the Moon carrying any ambient magnetic field. When the expanding gas converged on the antipode, it compressed and amplified the magnetic field at the antipode. One result of the localized lunar magnetic fields is low concentrations of hydrogen from the solar wind in the soil.

Lunar Prospector came to an end on July 31, 1999, when the spacecraft was successfully targeted into a crater at the south pole. An item carried by *Lunar Prospector* ensured the hearts of lunar enthusiasts went with the probe to the Moon. Secured within the spacecraft was a vial containing some of Eugene Shoemaker's ashes. Shoemaker had been cut down, still in his prime, by a tragic car crash in Australia the summer before Prospector launched to the Moon. Now the scientist who labored so hard to be the first geologist on the Moon finally got his ride to that natural satellite. When *Lunar Prospector* crashed into the Moon, Eugene Shoemaker became a part of it.

At the beginning of the 21st century, a new armada of spacecraft prepared to explore the Moon. These were not built in the U.S. or Russia and were not part of the now long-gone political Space Race. Instead, Europe and Asia set their sights on the Moon with innovative spacecraft based on the robust, proven hardware designs of commercial communications satellites. To get to the Moon, these craft would utilize low-energy trajectories that allowed lifting

The European Space Agency's SMART-1, launched on September 27, 1993, was western Europe's first Moon probe and the first Moon probe launched from South America.

heavy spacecraft with the use of modest, less costly rockets. Cape Canaveral and Baikonur would now be joined by launch sites in South America, India, Japan, and China as lunar departure points.

Continental Europe's only Moon mission to date (if you discount Russia, formerly a part of the USSR, as part of modern Europe) was SMART-1 (Small Missions for Advanced Research in Technology) flown by the European Space Agency (ESA). The spacecraft was a one-meter square structure with solar panels spanning 14 meters, and a launch mass of 366.5 kilograms. To reach the Moon, the craft used a low-energy, long-duration trajectory like that pioneered by Hiten and *Clementine*. Launched as a hitchhiker payload aboard an Ariane 5 rocket from Kourou, French Guiana, on September 27, 2003, SMART-1 was the first lunar launch from South America and carried 19 kilograms of instruments to study the evolution of the Moon, search for polar ice and investigate past lunar volcanic and tectonic activity.

To reach the Moon from its initial geosynchronous transfer orbit, SMART-1 used an experimental solar-electric propulsion system commonly known as an ion drive. With a thrust of 70 millinewtons, the engine was weak but could operate for thousands of hours before exhausting its 82-kilogram supply of supercritical xenon propellant. Over a period of 14 months, the ion drive gradually elongated SMART-1's orbit out to a lunar distance. By January 2005 the craft was in a 471-by-2,880-kilometer polar orbit and science operations began the next month. The experimental ion drive ran out of fuel in September 2005 after firing for almost 5,000 hours, but the science mission continued until August 2006. SMART-1 deliberately crashed into the Moon on September 3, 2006, near Lacus Excellentiae. Europe's first dedicated lunar mission was conducted for about $90 million.

Japan Commits to the Moon

Japan's 2,914-kilogram Kaguya orbiter carried a large suite of science instruments for its two-year mission and demonstrated Japan's commitment to advanced lunar exploration.

Fifteen years after Hiten impacted the surface, Japan returned to the Moon with the Selenological and Engineering Explorer, known as SELENE. The new Japanese lunar probe was a rectangular spacecraft measuring 2.1-by-2.1-by-4.8-meters with a launch mass of 3,020 kilograms and was the largest lunar spacecraft since the Apollo program in the 1970s.

SELENE was a poetic adaptation of an otherwise scientific name, but the Japanese public selected the name Kaguya for the lunar prince in the classic Japanese folk tale *The Tale of the Bamboo Cutter*. Kaguya carried twin one-meter-high, 53-kilogram subsatellites that were released in lunar orbit. Officially designated Vstar and Rstar, they were also given names from folklore, Okina and Ouna.

Kaguya was launched on September 14, 2007, from the Tanegashima Space Center on Tanegashima Island by a Japanese H-IIA launcher, a license-built variant of the American Delta rocket. From an initial elliptical orbit, Kaguya used its own thrusters to progressively raise its apogee until it entered lunar orbit on October 3, 2007. On October 9, 2007, the Okina radio relay subsatellite was released to maintain contact with Kaguya when it was behind the Moon.

An exciting part of the Kaguya mission was a high-definition TV camera provided for public outreach by the Japan Broadcasting Corporation. Over a 19-month period, the camera broadcast 1.3 terabytes of video, including fascinating scenes of the Earth rising above the lunar horizon.

This was followed on October 12 by the release of Ouna, a radio relay for very long base line interferometry to map the Moon's gravity field. On October 19, Kaguya reached its 100-kilometer circular science orbit and by December 21, 2007, all 15 observation experiments were operational.

In early 2009, five active spacecraft were in orbit around the Moon, none of them were from the classic U.S. or Soviet Space Race superpowers: one from China and India, and three from Japan. The 21st-century Moon reflected the changes in a post-Apollo world where newly named lunar landforms would have Asian names.

The Terrain Camera on the Japanese orbiter was an order of magnitude better than the high-resolution camera aboard *Clementine*. The cameras used what was called the pushbroom imaging principle. Fore- and aft-looking cameras swept a single row of horizontal pixels as the spacecraft's motion carried the field of view vertically across the lunar terrain. Resolution was 10 meters per pixel and the twin cameras scanned the same overlapping 35-kilometer-wide swath of terrain from a perspective separated by 60 kilometers. This allowed the creation of highly detailed three-dimensional images. Data from the Terrain Camera demonstrated that some of the mare-forming volcanism on the far side continued until 2.5 billion years ago, and detailed gravity maps of the far side were also generated.

A curiosity found by the Terrain Camera was a "halo" area around the Apollo 15 landing site where the lunar module *Falcon* landed near Hadley Rille. The halo, a brightening of the lunar surface that had been swept by the exhaust plume of *Falcon*'s descent engine, spanned about 200 meters. Pre-landing images taken from *Falcon* during descent to Hadley Rille did not show the light-colored smudge now visible in Kaguya's images.

The orbiter's laser altimeter gathered six million surface height measurements at latitudes up to 75 degrees, creating an improved 3D terrain map of the Moon. Kaguya's spectral imagery provided the strongest proof yet that the Moon was once covered by a global magma ocean. The spectral analysis showed the highland areas, especially around the rims of basins, were composed of almost pure anorthosite, an aluminum-rich mineral that had to have formed at the end of the Moon's accretion. The lack of iron and other dense minerals shows the aluminum rose to the top of a melted lunar surface while the heavier iron sank in the fluid magma.

In addition to instruments to study the origin and evolution of the Moon's geology, Kaguya had an experiment that made it popular with the public. Two separate color HDTV cameras periodically beamed an astronaut's eye view of the lunar surface as Kaguya passed over major landmarks. The cameras sent back between four and ten videos per month at a resolution better than most of the *Clementine* images taken the previous decade.

At the completion of its extended mission, Kaguya was deliberately crashed into the Moon on June 10, 2009, striking the rim of an unnamed crater southeast of the Gill crater. The HDTV signal continued until impact. The 3.9-meter Anglo-Australian Telescope recorded the infrared flash of Kaguya's impact.

Asia Becomes a Lunar Exploration Giant

Following the Asian tradition of choosing names from folklore, China's lunar spacecraft was named Chang'e (pronounced "Chang-uh") after the goddess who took a magic potion to fly to the Moon. Chang'e 1 was based on the domestic Chinese DFH-3 communications satellite bus. Half of the 2,350-kilogram craft's mass was thruster fuel to loft Chang'e out of Earth orbit and into lunar orbit. The box-shaped 2-by-1.7-by-2.2-meter craft carried 24 science instruments totaling 130 kilograms.

After launching on October 24, 2007, aboard a Long March 3A rocket, Chang'e 1 took a looping 15-day journey to reach the Moon, followed by a spiraling descent to its 200-kilometer science orbit on November 5.

To celebrate achieving lunar orbit, the spacecraft transmitted 30 popular classical Chinese songs including "My Motherland," "The Song of the Yangtze River," and "High Mountains and Flowing Waters." The pride China took in their first lunar mission was demonstrated by Chinese Premier Wen Jiabao personally displaying the mission's first lunar photos at the Beijing Aerospace Control Center.

Chang'e 1 successfully completed its 12-month mission and flew an additional four-month extended mission, transmitting 175 gigabytes of data before being deliberately crashed into northern Mare Fecunditatis on March 1, 2009. The following November, a map of the entire lunar surface surveyed by Chang'e was released. The map's similarity to *Clementine* images led some detractors to accuse China of purloining the *Clementine* data. Fortunately, many Western planetary experts debunked the jingoistic sensationalism and confirmed the Chinese lunar images were genuine.

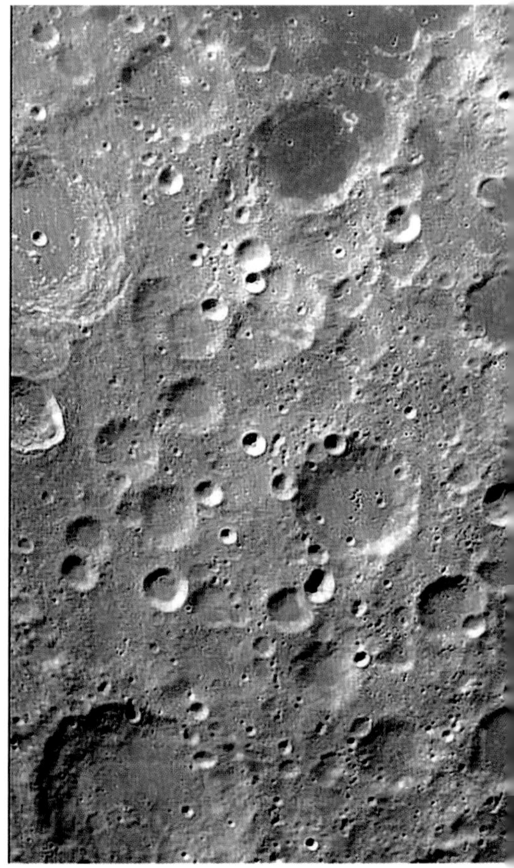

The pride that China took in its initial lunar mission was demonstrated by Chinese Premier Wen Jiabao personally unveiling the first lunar photos transmitted by Chang'e 1. The 19-image mosaic showed the territory around the Gill crater near the Moon's southeastern limb.

Since its launch in 2009, NASA's Lunar Reconnaissance Orbiter (LRO) continues to image the lunar surface looking for changes and fresh meteorite impacts.

The combined LRO and LCROSS mission was launched from Cape Canaveral, Florida, on June 18, 2009, aboard an Atlas V booster. The rocket had an evolved version of the American Atlas ICBM as the first stage and the Centaur rocket as a second stage. The Centaur had originally been developed in the early 1960s in part to launch the Surveyor program to the Moon. Now 42 years after the last Surveyor reached the lunar surface, the Centaur again was pressed into action for another lunar mission. action for another lunar mission.

Once in space, LRO separated from the Centaur booster while LCROSS remained attached to the spent rocket. After a 4.5-day journey from Earth, LRO entered lunar orbit while LCROSS and the Centaur rocket sped past and looped into a 37-day Earth orbit that intercepted the Moon's south pole three and a half months later.

LRO's complement of seven instruments were designed for two purposes, high-resolution imagery and the search for ice. One of the first targets for the LRO camera, which could resolve half a meter on the lunar surface, were the Apollo landing sites. The images were of historical importance and useful against growing legions of science deniers who claimed the Apollo landings were faked. But more important, a greater understanding of the landing site geology and topography will, even 40 years after the landings, enhance the science return from the Apollo explorations. The images were so detailed the shadow of the American flags planted by Apollos 12, 16, and 17 revealed that they were still upright after decades on the Moon. The Apollo 11 Tranquility Base image showed the individual footpads on the lunar module *Eagle*. The Apollo 12 site showed the lander *Intrepid*, the nearby Surveyor 3 unmanned lander, and Pete Conrad and Alan Bean's footprint trail between the two spacecraft. Apollo 14 images showed Alan Shepard and Stuart Roosa's footprints as they approached within a heartbreaking 15 meters of their target of Cone crater but could not recognize the terrain, thus missing the crater completely. Similar images of the Apollo 15, 16, and 17 sites were also returned. It was good to see the four-decade-old lunar modules still intact. Some engineers had feared the residual acidic hypergolic fuel in the lunar module's descent stage tanks might, over time, corrode through the tanks and explode as has happened in Earth-orbiting rocket bodies.

The Diviner radiometer aboard LRO found the coldest places in the solar system on the floor of some polar craters that never see sunlight. Areas at the south pole were as cold as -238 degrees Celsius and at the north pole near the southwestern edge of the Hermite crater, temperatures plunged to -258 degrees Celsius, colder than far away Pluto. These areas are cold enough that molecular water formed by interaction between the chemistry of the regolith and solar wind particles, as well as water gained from impacting comets, is frozen so solid that it will not sublime into the vacuum of space.

The Lunar Crater Observation and Sensing Satellite (LCROSS) separated from its Centaur rocket stage and trailed by four minutes when the Centaur impacted the Moon. Instruments aboard LCROSS detected water in the debris plume scattered by the Centaur impact within the Cabeus crater near the lunar south pole.

The Lunar Atmosphere and Dust Environment Explorer (LADEE) was a low-cost mission that filled many gaps in our understanding of the space environment near the Moon's surface.

The upcoming launch of the Chinese Chang'e 3 lunar landing mission initially caused concern among LADEE researchers because the Chinese craft's rocket exhaust could contaminate the pristine lunar environment and complicate the analysis of the rarified chemistry above the Moon. However, these concerns were unfounded. After Chang'e 3 entered lunar orbit on December 6, 2013, LADEE mission scientists discovered they could use the known chemistry of the rocket's exhaust to calibrate LADEE's instruments. Additionally, tracking the water created in the Chang'e 3 rocket exhaust provided insights about how molecular water migrates into polar cold traps.

LADEE's final orbit was lowered to within 300 meters of the surface on April 17, 2014. Helium, neon and argon were the most abundant gases detected. LADEE also detected increased amounts of lunar dust, including sodium, after the Geminid meteor shower in December 2013. Atomic oxygen, iron, titanium and components of lunar basalt, were also detected above the Moon. On April 18, 2014, LADEE struck the rim of crater Sundman 500 kilometers west of the Oceanus Procellarum shoreline leaving a V-shaped debris pattern that was photographed by NASA's Lunar Reconnaissance Orbiter.

China Lands on the Moon

China's first lunar landing mission, Chang'e 3, was launched on December 1, 2013, by a Long March 3B rocket. The population density around the Xichang Satellite Launch Center in southern Sichuan was so dense that prior to liftoff, 160,000 people had to be evacuated, and closer to the launch site, 20,000 people were moved to a school auditorium during the launch.

The 3,780-kilogram Chang'e 3 lander entered a 100-kilometer lunar orbit on December 6, 2013, and subsequently lowered to 15-by-100-kilometers prior

The LADEE mission launch by a Minotaur rocket from the coast of Virgina became an internet sensation when the silhouette of a frog appeared against the bright rocket exhaust. The fate of the frog blasted from the launch pad is unknown.

China's Chang'e 3 landed on Mare Imbrium on December 14, 2013, and deployed a 140-kilogram rover named *Yutu,* which translates from Mandarin as the Jade Rabbit. The rover carried 20 kilograms of instruments, including a ground penetrating radar that examined the surface soil to a depth of 30 meters and the structure of the lunar crust to a depth of several hundred meters.

to the December 14 landing. The craft descended to a hover at 100 meters altitude while the autonomous navigation system steered the vehicle away from obstacles. After descending to four meters, the engines cut off and Chang'e 3 dropped to the surface. China became the third nation to successfully land on the Moon and Chang'e 3 was the first craft to land on the Moon since the Soviet Luna 24 sample return mission 37 years before.

Initially, the landing was to be within Sinus Iridum, a 400-kilometer-wide horseshoe-shaped bay on northwest Mare Imbrium. When the craft came to rest on Mare Imbrium 135 kilometers southeast of Promontorium Laplace, no explanation was given about the landing site change.

A 140-kilogram 1.5-meter-long, one-meter-high rover named *Yutu* was deployed onto the surface. The rover's name translates as the Jade Rabbit. The rover carried a ground-penetrating radar to study the subsurface structure to a depth of several hundred meters and spectrometers to analyze the chemistry of the soil. Although designed for a 90-day lifetime, the Jade Rabbit returned data until March 2015. Unfortunately, the rover was unable to move after the second lunar night and only traveled a total of 114 meters.

An astronomical instrument aboard the lander was the Lunar-based Ultraviolet Telescope, or LUT. This 50-mm Ritchey–Chrétien telescope studied deep space objects in the near ultraviolet spectrum, taking advantage of the slow lunar rotation to allow extended exposures of objects as dim as magnitude 13. The LUT was still in operation seven years after landing.

To honor the Chinese lunar landing, the International Astronomical Union approved the landing site name of Guang Han Gong, which translates from Mandarin as "extensively cold palace." The Chang'e 3 lander was still in contact with Earth 10 years after touchdown.

China's Chang'e 4 made the first touchdown on the Moon's far side on January 3, 2019, and deployed the *Yutu-2* rover which explored the interior of the Von Kármán crater.

The next two Chinese lunar missions were in support of upcoming landings on the far side and to prove the technologies developed for a sample return mission. The 3,380-kilogram Chang'e 5-T1 spacecraft launched on October 23, 2014, aboard a Long March 3C rocket and looped around the Moon 97 hours later. As the craft returned to Earth, a 335-kilogram bell-shaped lunar sample reentry vehicle separated and tested the craft's ability to withstand the high velocity return from the Moon. The craft, looking like a scale model of the Chinese Shenzhou manned spacecraft, successfully landed in Mongolia 196 hours after launch while the service module remained in Earth orbit.

The Chang'e 5-T1 service module, based on the design of the previous Chang'e 2 orbiter, was maneuvered into an elliptical lunar orbit on January 13, 2015. In February and March, the craft performed a "virtual rendezvous" in lunar orbit to practice procedures to be used in the Chang'e 5 sample return mission. The future Chang'e 5 landing site was also photographed in detail. The third stage rocket, circling Earth for over seven years, stuck the far side of the Moon within the Hertzsprung crater on March 2, 2022.

To communicate with the soon-to-be-launched Chang'e 4 lander on the far side of the Moon, the Queqiao-1 relay satellite was launched by a Long March 4C on May 20, 2018. On June 14, the craft was maneuvered into the Lagrange L2 point opposite the Moon, where it would remain in a stable location to receive signals from the far side and relay them to Earth. Queqiao translates as the "Magpie Bridge." The satellite utilizes a 4.2-meter folding dish antenna for both signal relay and radio astronomy experiments. China entered into an agreement with the United States allowing NASA to also use Queqiao-1 on future lunar missions.

Using the experience gained from the successful Chang'e 3 landing, the backup spacecraft was modified and flown as Chang'e 4, the first craft to land on the far side of the Moon. The *Yutu-2* rover was upgraded to avoid the mobility issues encountered by the first *Yutu*.

Chang'e 4 launched aboard a Long March 3B rocket on December 7, 2018, and entered lunar orbit on December 12. The orbit was lowered to 15 kilometers on December 30 in preparation for landing on January 3, 2019. Twelve hours after landing in the Von Kármán crater within the massive South Pole–Aitken basin on the far side, the *Yutu-2* rover was deployed on the surface. Contact with Earth was maintained through the Queqiao-1 relay satellite. The modifications to both spacecraft were productive as both the Chang'e 4 lander and *Yutu-2* were still operational in May, 2024 and the rover had traveled one-and-one-half kilometers. *Yutu-2* carried spectrometers and ground penetrating radar. In May, 2019, the rover identified rocks that originated from deep in the Moon's mantle.

A fascinating experiment carried aboard the Chang'e 4 lander was a three-kilogram sealed Lunar Micro Ecosystem biosphere containing cottonseed, rapeseed, potato seeds, yeast, and fruit fly eggs. After landing, the seeds were watered and photos showed they had sprouted on January 15, 2019. Unfortunately, the 100-day growth experiment was prematurely terminated a day later when temperature control was lost and the seedlings froze.

The Chang'e 4 landing site within Von Kármán holds both scientific and symbolic value for the Chinese. Theodore von Kármán was the PhD advisor for Qian Xuesen, a leading American rocket researcher in the 1940s. After being falsely accused of being a Communist sympathizer, Xuesen was deported in 1955. Returning to home, he became the head of the missile and space technology development in China.

Reinforcing the connection between von Kármán and the Chinese space program, the landing site within the Von Kármán crater was officially designated Statio Tianhe. The name Tianhe was an ancient Chinese name for the Milky Way. At an elevation of 5,935 meters below mean lunar elevation, Statio Tianhe is the lowest elevation of any landing site to date. The Von Kármán crater's central peak was named Mons Tai, after the terrestrial Mount Tai in Shandong, China. Additionally, three craters near the landing site, between two to four kilometers wide, were named Zhinyu, Hegu, and Tianjin after ancient Chinese constellations.

An International Fleet of Explorers

Israel became the seventh nation to reach the Moon, if you count the combined nations under the ESA banner as one entity, but their soft-landing attempt failed. The *Beresheet* lander was privately developed by Israel Aerospace Industries and the education advocacy group SpaceIL and launched aboard a Falcon 9 rocket from Cape Canaveral, on February 22, 2019. Originally named the *Sparrow*, the lunar lander was renamed *Beresheet*, which is Hebrew for "In the beginning." SpaceIL promoted the mission with the theme "small nation, big dreams." The 150-kilogram, two-meter-wide, 1.5-meter-high spacecraft performed a series of Earth-orbit raising maneuvers to reach the Moon on April 4, 2019. A landing attempt on Mare Tranquillitatis on April 11, 2019, was unsuccessful when one of the spacecraft's accelerometers failed and an improper recovery command sequence shut down the craft's engine. *Beresheet* was subsequently destroyed when it impacted the Moon at 3,000 kilometers per hour.

India returned to the Moon when its Chandrayaan-2 lunar landing mission departed Earth on July 22, 2019. An Indian-developed LVM3 M1 rocket boosted the craft into an initial elliptical Earth orbit and the craft slowly raised its apogee over the next 22 days to reach the Moon. The 3,850-kilogram spacecraft entered lunar orbit on August 20, 2019. The Chandrayaan-2 orbiter carried a 1,471-kilogram lander named *Vikram*, Sanskrit for "valor" and a 27-kilogram rover named *Pragyan*, Sanskrit for "wisdom." *Vikram* was designed to land on a slope of up to 12 degrees.

Chandrayaan-2's orbit was lunar circularized by September 1 and the *Vikram* lander separated on September 3. After lowering its orbit to 35-by-101 kilometers, *Vikram* attempted to land near the south pole on September 6. Initial descent was good, but a software flaw resulted in loss of control close

The privately funded Israeli *Beresheet* lunar lander was a bold attempt by private industry and the space outreach group SpaceIL to land a robotic spacecraft on the Moon. Unfortunately, control was lost during descent to the surface and the craft crashed.

to the Moon and the craft was destroyed on impact within 500 meters of its target landing site.

Although the *Vikram* landing failed, the instrument-laden orbiter remained in operation for five more years. The orbiter's cameras were capable of a one-meter resolution and returned the highest resolution images of the Moon to date. An exciting find was the Moon's sodium tail, detected by the orbiter's X-ray spectrometer. Chandrayaan-2 demonstrated that sodium atoms dislodged from the surface by the solar wind and meteorite impacts are accelerated away from the Sun by the pressure of sunlight and create a sodium tail pointing away from the Sun.

The third decade of the 21st century saw a flurry of international lunar missions reminiscent of the 1960s race to the Moon. First up was Chang'e 5 which advanced China's lunar explorations by performing the first sample return mission since the Soviet Luna 24 in 1976.

The Chang'e 5 orbiter, descent module, and ascender used to return the lunar sample had a combined mass of 8,200 kilograms. After launch aboard a Long March 5 rocket on November 23, 2020, the craft reached lunar orbit on November 28. The 3,200-kilogram lander detached and touched down on December 1. The lander was equipped with both a drill and scooping device that collected 1.731 kilograms of soil and rock from the point farthest north yet sampled, a location on northern Oceanus Procellarum about 130 kilometers

The Chang'e 5 sample return capsule landed on the snow-covered steppes of Mongolia on December 16, 2020, marking China as the third nation to return a sample of the lunar surface to the Earth.

northeast of Mons Rümker. A ground penetrating radar analyzed the rock layers under the landing site. On December 3, the 700-kilogram ascender lifted off and two days later performed an automatic rendezvous with the orbiter. The landing spacecraft was not expected to, and did not, survive the ascender's liftoff, and was thus not equipped with a radioisotopic heater to allow surviving the cold lunar night. The lunar sample canister was transferred into the 300-kilogram Earth return capsule. On December 7, the ascender deliberately crashed into the Moon. On December 11, the orbiter boosted the return capsule into an Earth return trajectory. On December 16, 2020, the return capsule landed on snow-covered Inner Mongolia.

A significant find from laboratory analysis of the Chang'e 5 soil samples was laboratory proof of active volcanism on the Moon only 125 million years ago. This finding erases the previous belief that the Moon has been volcanically dormant for the past billion years and shows the Moon has been volcanically active for far longer than previous conclusions suggested. Another significant find was hydroxyl molecules, or one atom of hydrogen bound to one atom of oxygen, at concentrations of up to 120 parts per million.

The next lunar mission was the Cislunar Autonomous Positioning System Technology Operation and Navigation Experiment, known as CAPSTONE. This mission was the cheapest lunar mission yet flown and the first launch toward the Moon from New Zealand. The 25-kilogram CAPSTONE spacecraft was developed by private industry for $13.7 million and launched aboard a Rocket Lab Electron rocket for $9.93 million on June 28, 2022. After employing a lengthy but fuel-efficient ballistic trajectory, CAPSTONE entered a 1,500 by 70,000-kilometer polar orbit around the Moon on November 14, 2022. The mission objectives were to explore the stability of an elliptical polar lunar orbit for the proposed, but now canceled, Lunar Gateway space station planned as a waystation for later Artemis manned lunar missions. By 2026, CAPSTONE was still operating in its far-reaching orbit.

Korea joined the expanding list of nations exploring the Moon when the Korea Pathfinder Lunar Orbiter (KPLO) was launched aboard a Falcon 9 rocket from Cape Canaveral on August 4, 2022. A Korean national contest to name the spacecraft resulted in the popular name Danuri, the combination of the Korean words for "Moon" and "enjoy." The 678-kilogram spacecraft flew a 135-day ballistic trajectory and arrived in lunar orbit on December 7, 2022. By December 28, the craft was maneuvered into its 100-kilometer polar science orbit.

In addition to multiple cameras, a magnetometer and a gamma ray spectrometer to determine surface composition, Danuri carried a NASA-funded instrument called ShadowCam. This hypersensitive photographic instrument was designed to peer into permanently shadowed polar craters to search for signs of ice. ShadowCam imaged the normally dark regions by sunlight reflected off the crater rim or by sunlight reflected from Earth. The instrument was 200 times more sensitive than the camera aboard NASA's Lunar Reconnaissance Orbiter and could resolve 1.7 meters per pixel. The spacecraft and ShadowCam were still operational in 2025.

One of the ShadowCam's achievements was photographing the Lunar Reconnaissance Orbiter spacecraft as it passed 18 kilometers below Danuri at a velocity 3,180 meters per second relative to Danuri. The resulting image was smeared by the high velocity, but the LRO spacecraft was recognizable.

After decades of mission planning and program cancellations by presidential administrations, the American manned return to the Moon under NASA's Artemis program finally achieved its first test flight on November 16, 2022. NASA's Space Launch System (SLS) rocket succeeded in propelling an unmanned Orion spacecraft and its European-developed service module into a looping 25-day trajectory that circled the Moon twice and reached a distance of 452,210 kilometers from Earth.

During its 2.1-million-kilometer journey, the Artemis I mission released 10 subsatellites from the U.S., Italy, and Japan that were to study the lunar environment. Unfortunately, five of the subsatellites failed to complete their mission. Orion looped within 128 kilometers of the Moon before returning to Earth and the spacecraft was recovered after splashdown off the coast of Baja California, Mexico. Although the Orion performed well in space, remedying the unexpected excess erosion of its heat shield upon entry into atmosphere delayed the manned flight of Artemis II around the Moon until 2026.

India's domestically developed Launch Vehicle Mark-3 (LVM3) launched the Chandrayaan-3 mission on July 14, 2023. The successful landing of the *Vikram* lunar lander on August 23, 2023, marked India as the fourth nation to successfully land on the Moon.

Thirteen years after its previous lunar mission, Japan attempted a soft landing with the privately financed HAKUTO-R mission. The 2.3-meter-tall, 1,000-kilogram lander carried two small rovers, the SORA-Q from Japan and the *Rashid* rover provided by the United Arab Emirates. HAKUTO-R launched from Cape Canaveral aboard a Falcon 9 rocket on December 11, 2022, and was to attempt a landing in the Atlas crater near the Moon's northeastern limb. After traveling 1,400,000 kilometers, the craft entered lunar orbit on March 23, 2023. A landing attempt on April 25 failed when a software glitch misinterpreted the radar altitude data and held HAKUTO-R at a hover 5 kilometers above the Moon until fuel depletion. Although the craft crashed, the onboard *Rashid* rover made the UAE the ninth nation to impact the Moon.

India became the fourth nation to soft land on the Moon when the Chandrayaan-3 mission delivered the *Vikram* lander to a safe landing near the south pole. The 3,900-kilogram Chandrayaan-3 was similar in form to Chandrayaan-2 but was upgraded to prevent the issues which caused the previous landing attempt to fail. Launched aboard an Indian LVM3 rocket on July 14, 2023, the craft entered lunar orbit on August 5. A landing on August 23, 2023, between the Manzinus and Boguslawsky craters was followed by the deployment of the *Pragyan* rover. Over the course of the lunar day, *Pragyan* traveled 103 meters and analyzed 23 locations with its alpha particle X-ray spectrometer. Results of these studies showed the widespread presence of anorthositic silicate rock, or a rock composed primarily of one type of mineral. Anorthosite is thought to have been created on the Moon when it was still molten after its formation and the lightweight minerals like silica floated to the top of a global magma ocean. Similar anorthosite deposits have been found in Apollo and Luna samples collected in the equatorial regions of the Moon. The presence of anorthosite close to the south pole reinforced the theory that the

India's *Vikram* lander was imaged on the lunar surface by the *Pragyan* rover, which explored the region for a lunar day.

Several hours after *Vikram*'s touchdown, the *Pragyan* rover rolled down ramps onto the surface to explore the local geology.

Initial analysis of the far side samples revealed a density of 3.035 grams per cubic centimeter, looser and more porous than samples recovered from the nearside. This finding is consistent with the sample being recovered from the crushed interior of the Apollo Basin, which itself lies inside the larger South Pole–Aitken basin. The samples were composed of 45.6% silicon dioxide, 14% aluminum oxide, and 12% calcium dioxide. Most of the sample was considered local basalt, with particles of ejecta from different locations mixed in. With the far side samples considered, Chinese lunar specialists stated the Moon's surface can be described in three categories: the Procellarum KREEP Terrane containing a higher concentration of rare earth and radioactive elements, the Feldspathic Highlands Terrane representing the original crust of the Moon, and the South Pole–Aitken Terrane representing the impact-basin-modified far-side southern hemisphere.

Fifty-three years after the last manned Apollo mission returned to Earth, the Moon still proves to be a harsh mistress. As the first quarter of the 21st century ended, international lunar exploration continued to suffer both administrative and physical setbacks. More than 50 years after the glory of Apollo, flights to the Moon have still not matured into routine success. Attempts to land robotic science instruments on the Moon often meet with either the fatal swipe of a budget cutter's pen, or a dismal crash on the harsh lunar surface.

Questionable fiscal decisions led to the July 2024 cancellation of the already completed Volatiles Investigating Polar Exploration Rover (VIPER), upon which great hopes lay to ferret out ice deposits near the lunar south pole. The completed VIPER and its suite of science instruments had been developed at a cost of $450 million. Then as the craft was ready for its flight to the Moon, NASA canceled the program to save a paltry $84 million. Even more shocking to the science community was NASA's decision to fly the Griffin-1 lunar lander, which was to deliver the VIPER to the lunar south pole, with no science payload at all. Initially, a "mass simulator" producing no lunar science at all was to be flown to the Moon in 2026 at a cost of $323 million.

In the wake of the scientific outcry about VIPER's cancellation, Astrobotic Technology, the developer of the Griffin-1 lander, now plans to deliver its own half-metric-ton FLEX Lunar Investigation Platform (FLIP) rover, carrying 30 kilograms of instruments, to the lunar surface in 2026.

Following the loss of half of the Artemis I subsatellites, Japan's HAKUTO-R, Russia's Luna 25, and the *Peregrine* and *Odysseus* landers, robotic lunar exploration received a welcome boost with the fantastic success of the inaugural Blue Ghost Mission 1. Following its January 15, 2025, launch aboard the same Falcon 9 rocket that lofted Japan's second HAKUTO-R, the mission was playfully named "Ghost Riders in the Sky" after the popular 1970 Johnny Cash song. The spacecraft completed a perfect March 2, 2025, landing on Mare Crisium. Built by Texas-based Firefly Aerospace, the gold-foil-covered 6.6-foot-high lander spanned 11.5 feet across its landing legs and had a mass of 1,517 kilograms (3,344 pounds) when fully fueled. It was the first successful private lunar mission funded under NASA's CLPS program. For good luck, the lander's frame was signed by Apollo Moon walkers Buzz Aldrin and Harrison

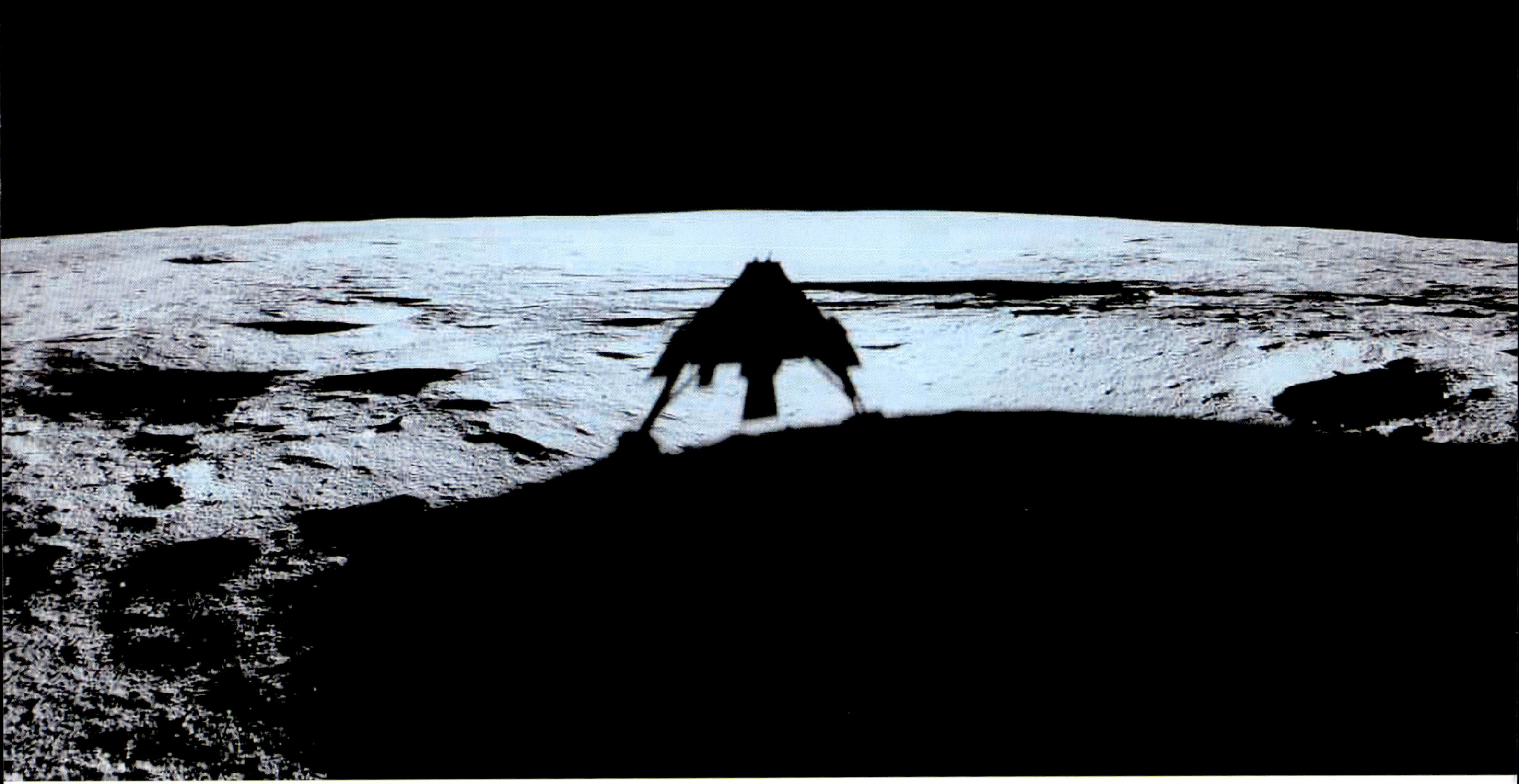

Schmitt. Designed to operate for one lunar day, the Blue Ghost ceased operation after local sundown on March 16, 2025.

A video released by Firefly Aerospace showed the lander's shadow, starting at 28 meters (91 feet) altitude, as it slowly descended to the surface. Soon, the shadow of an astonishingly violent swirling dust storm whipped up by rocket exhaust turned into a horizon-obscuring haze. With shocking swiftness after the engine shut down, the dust settled quickly in the airless lunar environment to reveal a stone hurled toward the horizon and splash into the dust to the left of the lander's shadow.

Upon touchdown, the first of the 10 lunar surface experiments was a success. NASA's Langley Research Center provided the Stereo Cameras for Lunar Plume-Surface Studies (SCALPSS), a six-camera suite that recorded 3,000 frames studying how rocket exhaust reacted with the loose lunar dust and regolith. Even after the Apollo manned landings half a century ago, the dynamics of rocket exhaust impinging on the regolith are not fully understood. The SCALPSS experiment is thus vital to assist the planned Artemis manned landings later this

The Blue Ghost became the first successful commercial mission to the Moon when it landed safely on Mare Crisium on March 2, 2025, and transmitted an image of its own shadow with the Earth hanging above the lunar horizon.

decade. Unlike Apollo, which used a different engine to take off from the Moon after landing, the SpaceX Starship HLS will use the same engines for landing and takeoff, and they must not be damaged by debris whipped up during landing.

Of particular interest was Blue Ghost's Lunar Magnetotelluric Sounder (LMS) experiment. Magnetotellurics have been used on Earth for 50 years for finding oil, water, geothermal and mineral resources as well as to understand geologic processes, but this was the first application of such technology on the Moon. Magnetotellurics use the natural variation in surface electrical and magnetic fields to calculate how easily electricity flows through subsurface materials. Shortly after landing, the experiment deployed four sensors, each more than 60 feet away from the Blue Ghost lander at 90-degree angles, to characterize the lunar subsurface. A delightful video released by Firefly showed the deployment of the softball-sized magnetotelluric sensors. Like an outfielder hurling the game-winning ball toward home plate, the Blue Ghost heaved the sensors and their trailing electrical cable far out from the lander. The LMS experiment is expected to provide insights into the Moon's mineral differentiation and thermal history to a depth of 1,120 kilometers, about two-thirds of the Moon's radius.

Another fascinating experiment provided by the Italian Space Agency was the Lunar GNSS Receiver Experiment (LuGRE) that investigated the use of GPS satellite signals to navigate on the Moon. The LuGRE receiver was able to pick up GPS signals both on the way to the Moon and from the lunar surface. This shows that the same GPS signals that guide us through our smartphones can also help future lunar explorers navigate the Moon's surface.

Unfortunately, the good luck of the Blue Ghost mission did not rub off on Japan's second HAKUTO-R lunar lander. Launched on January 15, 2025, by the same Falcon 9 rocket that lofted the Blue Ghost, the new Japanese lander was a near copy of the previous HAKUTO-R spacecraft, this time named *Resilience*. A five-kilogram lunar rover named *Tenacious* was among the experiments carried. Taking a longer, more energy-efficient route, *Resilience* entered lunar orbit on May 6, 2025, and attempted a landing on Mare Frigoris in the Moon's northern polar latitudes on June 5. Contact was lost with the lander shortly before the planned touchdown time. Analysis by SpaceIL, the craft's builder, points to a fault in the laser altimeter, leading to a hard landing and subsequent loss of the spacecraft.

On February 27, 2025, Intuitive Machines launched its second lunar lander aboard a Falcon 9 from Cape Canaveral. Similar in form to the company's first lunar attempt with the *Odysseus* lander one year before, the new mission, christened *Athena*, attempted a landing near Mons Mouton near the lunar south pole on March 6, 2025.

In a heartbreaking repeat of the company's previous landing which tipped over and crashed due to excess horizontal velocity, *Athena* also stumbled upon landing and tipped over onto its side. Contact was temporarily lost when one of the two antennas was damaged in the crash, but limited contact was reestablished for one day before the craft lost power and went silent. An iconic image relayed by *Athena* showed two landing legs pointed upward and bracketing a crescent Earth near the lunar horizon. Analysis of *Athena*'s crash pointed to a plume of lunar dust kicked up by rocket exhaust interfering with the laser altimeter, leading to a hard touchdown and toppling over.

Among its suite of experiments, *Athena* carried three rovers, none of which could be deployed, and the Polar Resources Ice Mining Experiment (PRIME) which was a copy of the ice prospecting drill that was to be carried by the canceled VIPER rover. The loss of the PRIME drill was a bitter blow to lunar scientists, still stung by NASA's strange decision to cancel the completed VIPER rover.

Launched as a secondary payload on the same Falcon 9 rocket that sent *Athena* to the Moon, the Lunar Trailblazer was intended to be a 210-kilogram lunar satellite to study the Moon's water. Sadly, the bad luck that dogged *Athena* to the lunar surface also blighted the Lunar Trailblazer as contact with the spacecraft was lost shortly after launch. It is assumed the silent craft flew past the Moon and did not enter lunar orbit. Attempts to contact Lunar Trailblazer were abandoned in July, 2025 and the mission was declared lost.

In a sad repeat of their failed first lunar landing, Intuitive Machines' *Athena* lander also toppled on its side upon touchdown. Before the craft lost power, it sent a photo of the Earth between the craft's upraised landing legs.

An Understanding of the Moon

Midway through the 2020s, the pace of international lunar exploration continues to accelerate, leading up to a manned return to the Moon with NASA's Artemis program and the Chinese space program later in the decade. But the efforts to return to the Moon in the 21st century lack the sparkle and glamor of the race to the Moon half a century ago. The U.S., India, Japan, and China have ambitious plans to send more lunar explorers to the Moon in the coming years. Today's lunar missions benefited from the pioneering 20th century explorations by refining the exploratory questions being asked and using advanced sensors and communications equipment not available in the heyday of Apollo. The result has been a bonanza of new lunar data that reveals a world more complex than we dreamed at the beginning of the Space Age.

Mountains of research papers and hundreds of magazine articles and books (including two by this author) have been written explaining the physics, chemistry, and joy of observing the Moon. But what is the average person's basic understanding about the Moon from seven decades of lunar research? After three quarters of a century of Space Age lunar explorations, this author sees three main questions that dominate our knowledge of the Moon.

The first is the question used to scientifically justify the massive expense of exploring the Moon in the 1960s: How did the Moon form? Strangely that may be the last question answered with certainty as lunar explorations continue in the 21st century. After sifting and digesting the masses of geologic data returned during the golden years of our space program, a consensus grew among scientists in the 1970s and 1980s for a theory that explains the formation of the Moon. Two research groups, one with William Ward and Alastair Cameron, and the other with William K. Hartmann and Donald R. Davis, developed the impact model of lunar formation. This theory supposes that 4.5 billion years ago the inner solar system was an immensely crowded place and a Mars-sized planet, unofficially known as Theia, was vying with Earth for orbital supremacy near the Sun. Eventually, smaller Theia struck Earth a glancing blow that shattered Theia and ripped off a major portion of the young Earth's mantle. Material from both planets combined into a blend of molten and collision-vaporized material. The iron and heavy elements in the huge rotating debris cloud coalesced to form the core of the new Earth while most of the lighter rocky silicate debris ripped off Earth's mantle remained in a Saturn-like ring orbiting close to the remains of Earth. Within one to 100 years, the ring material gathered into a ball the size of the Moon today. This planetary collision theory popularized in the 1980s, affectionately called the "Big Whack" model, is not universally accepted, but it remains the foremost contender because it answers the riddles of lunar formation better than other theories.

The second fundamental question is whether the Moon was molten early in its evolution. The answer to this has profound implications about the interior structure of the Moon and subsequent sources of the volcanism that modified the Moon's impact-battered surface. Data gathered from all over the lunar globe from the Apollo era to modern times points to the extensive spread of anorthosite across the Moon's face. The only source for the global presence of this lightweight mineral is that it floated to the surface of a global magma ocean shortly after the Moon's formation about four-and-a-half billion years ago.

In the mid-20th century, Harold Urey passionately presented his case that the Moon formed cold and represents a geologic time capsule of the early solar system. While Urey's "cold Moon" theory is now disproved by evidence of the global magma ocean, the Moon does indeed represent a geologic time capsule of how the Earth–Moon system formed. The Moon efficiently recorded the impact record of the early solar system. It was not until after the Apollo expeditions that NASA's planetary explorers revealed that all the rocky planets from Mercury to Mars, and the satellites of the outer planets, were pummeled by waves of vicious impacts nearly four billion years ago. We now understand that everything we see on the face of the Moon was created either by an impact or the subsequent volcanic modification of an impact feature. Our Moon preserved this impact record and displays it before our telescopic eyes, but it remained misunderstood until we ventured to the other planets after conquering the Moon. Only Earth, with its weather, oceans, and tectonic activity, has healed the collision wounds common to all solid bodies in the solar system.

The question being answered by modern lunar exploration is the amazing finding of water on the Moon. Its presence is counterintuitive to the science results of Apollo and its robotic preludes but nonetheless has repeatedly revealed itself in more refined 21st century explorations. Water is a critical resource for the future scientific inhabitation of the Moon, and its presence is a true game changer in how we can establish manned bases on that world.

Today, China stands as a lunar powerhouse, successfully landing four consecutive spacecraft on the Moon and demonstrating a robust space program capable of landing Chinese astronauts on the Moon by 2030. The Moon has lured human attention since humankind became a thinking, reasoning species and despite setbacks, spacefaring nations will persevere in their pursuit of its secrets.

Thanks to the international fleet of 21st century unmanned lunar probes, we now have a much better understanding of the chemistry and evolution of the Moon than we did when Apollo 17 lifted off from the Taurus–Littrow valley in 1972. But the question used to scientifically justify the expensive reach for the Moon in the 1960s remains unanswered. Despite half a century of probes sent by the U.S., Russia, Europe, India, Japan, and China, the definitive answer to how the Moon was created still eludes us. Intriguing theories dance before us, but there is still no universally accepted answer. Perhaps that is just as well, for it is unanswered questions that will keep luring us back to the dusty surface of our Moon, and human curiosity will perpetuate our quest to understand the universe and our place in it.

Understanding the Face of the Moon

The following images, all taken by the author from his personal observatory in central Texas, are a sampling of the exotic lunar terrain that awaits the next astronauts.

After thousands of generations of humans looking up at the Moon in the night sky, it has become an unconscious part of the human psyche. It has been worshiped as a deity, evoked a sense of fear among the superstitious, wonder among the curious, and romance in the lighthearted. The Moon spurred generations of prose and poetry that tugged at the human heart. The Moon is interwoven with the social and scientific fabric of mankind and is embedded in the culture, literature, and lore of civilizations all over the world. The Moon is part of humanity's inherent curiosity.

For the terrestrial observer, the Moon is the most accessible celestial object. The Moon laughs at light pollution and can be observed from the heart of the largest and brightest city. More detail can be seen on the Moon with the naked eye than can be discerned in a galaxy through a high-power telescope. A pair of binoculars will reveal more detail on the Moon than can be seen on the planets with the best telescope.

Viewing the Moon through a telescope reveals two diverse landscapes: one dark and flat, the other bright and heavily cratered or crossed by mountain chains. This telescopic appearance leads to the conclusion that these areas have two different compositions. Indeed, the dark regions were covered by dark lava flows from ancient volcanic eruptions while the brighter areas display the higher elevations of the original ancient surface of the Moon.

Navigating the face of the Moon is easy. The broad dark maria and landmark craters like Tycho, Copernicus and Plato allow a novice to "crater-hop" across the Moon, like the star patterns of the constellations allow us to star-hop to deep sky targets.

What do we see when we look at the Moon through a telescope? The answer from a novice viewer is an overwhelming number of craters, each different; like no two snowflakes being the same. At the same time, it is easy for the novice to overlook the obvious large-scale clues about the Moon's evolution that lie in plain sight — the huge impact basins now masquerading as the lunar maria. These are the most important clue about how the face of the Moon was created by massive asteroid impacts almost four billion years ago.

Once we get past the impression of a sandblasted, pockmarked Moon, we can recognize a remarkably diverse geologic surface. There is incredible complexity in the variety of lunar features, but a soothing simplicity in how they came into being. All lunar features, both large and small, were created by only two forces. Everything we see on the Moon originated with an impact, whether from a meteor, a comet or an asteroid. The other modifying force, volcanism, was often released by impacts.

Understanding how each of these world-building forces formed the face of the Moon allows us to have a greater appreciation of what we see in the telescope eyepiece: younger craters overlay the older craters, craters with rays are less than a billion years old, large craters near the edge of the volcanically formed dark maria are likely to also be volcanically modified after their creation. Similarly, the volcanic modification of the Moon can be traced by the lava flows covering basins and territories below mean lunar elevation, and the presence of lengthy fractures in the surface that were split open by volcanic pressures.

Two conclusions were drawn about the Moon before spacecraft began probing its surface. The first was the realization that the Moon underwent a terrific meteoric bombardment about half a billion years after it formed. During this period, the surface layer of the Moon was crushed and mixed by uncountable meteorite impacts. The other conclusion was William Hartmann's deduction that volcanic activity on the Moon was prevalent for almost a billion years beginning about 700 million years after the Moon's formation. These two activities, asteroid impacts and subsequent volcanic modification, formed the face of the Moon we see today.

Today we know the Moon is far from the pristine primordial object it was thought to be at the dawn of the Space Age. In the 1950s, it was assumed the Moon was a preserved piece of the early solar system that was akin to a 4.5-billion-year-old geologic time capsule. But this is not the case. The Moon is a rocky body that has been modified over the eons by both the exogenic forces of impacts, and the endogenic forces of volcanism.

After 60 years of Space Age research many unexpected geological facts have been discovered about the Moon, such as the amazing find that the Moon hosts large quantities of water ice, a resource critical to future habitation of the Moon. However, to date, it is apparent that the Moon has been, and remains, lifeless. Within the limited lunar samples returned to Earth, no living organisms, fossils, or organic compounds have been found. Far from the exaggerated sharp angular terrain of historical illustrations, lunar mountains are rounded and have shallow inclines, old craters are smoothed and rounded, and exposed rocks are worn by the timeless rain of micrometeorites and gaseous eruptions from the Sun.

All we see on the Moon, from the smallest rock to the largest mare, were not only created by impact but continue to be modified throughout time. For the telescopic observer, the legacy of the Moon's geologic evolution is a mesmerizing view of a world that is rugged yet serene, harsh yet beautiful, alien yet appealing, tortured yet calm, and strange yet friendly.

Both American and Chinese programs will soon land on the Moon and continue the explorations started during the stunning Apollo era half a century ago. If the recent finding of water on the Moon is an indication, the gush of new science from the Moon will provide answers to questions we do not yet know to ask. The future of lunar exploration is unfolding before our eyes. Prepare for the adventure of a lifetime as we virtually ride along with the next lunar explorers!

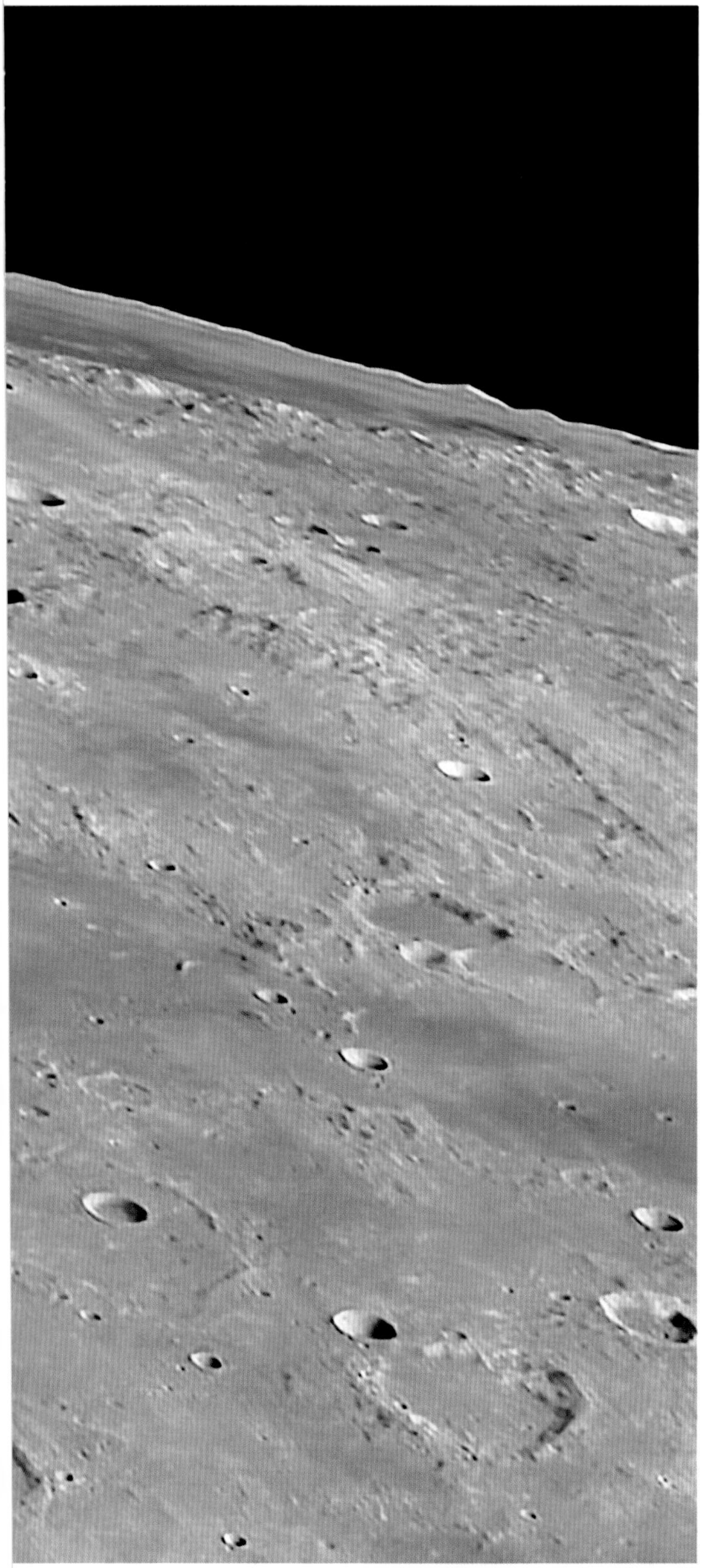

Endymion

Our oblique view of the lunar strongmen, Hercules and Atlas craters at the lower left, dark floored Endymion crater at the center, and the wide expanse of Mare Humboldtianum on the horizon at the upper right, unleashes the imagination by appearing like a view out the window of a spacecraft orbiting the Moon. Our slant view toward the horizon reveals the depth and texture of the lunar surface as craters and mountains rise up to greet our overhead passage.

Hercules spans 77 kilometers and displays a partially basalt-flooded floor punctuated by the impact of Hercules G. Older Atlas to the east spans 80 kilometers and is a floor-fractured crater where lavas welled up through fractures in the crater's floor then cooled, shrank, and created a network of rilles across its interior.

Closer to the horizon, the interior span of 126-kilometer-wide Endymion crater is completely obscured by basalt deposits. The dark floor of Endymion lies three kilometers below the mean lunar elevation and is remarkably flat, varying no more than 50 meters across the middle of the basalt field. The crater's interior is marred by relatively few small craters that formed after volcanism modified Endymion's interior.

On the horizon, the flat basalt plains of 231-kilometer-wide Mare Humboldtianum hug the Moon's northeastern limb. This dark spot was designated Zoroaster by Ricciolli in the 17th century. Almost two centuries later, the German lunar cartographer Johann Mädler named the region Mare Humboldtianum, or Humboldt's Sea, after his German contemporary, the explorer and naturalist Alexander von Humboldt. The overlaying 215-kilometer crater Bel'kovich, seen as the broad depression to the upper left of the darker maria, is often misidentified as part of Humboldtianum.

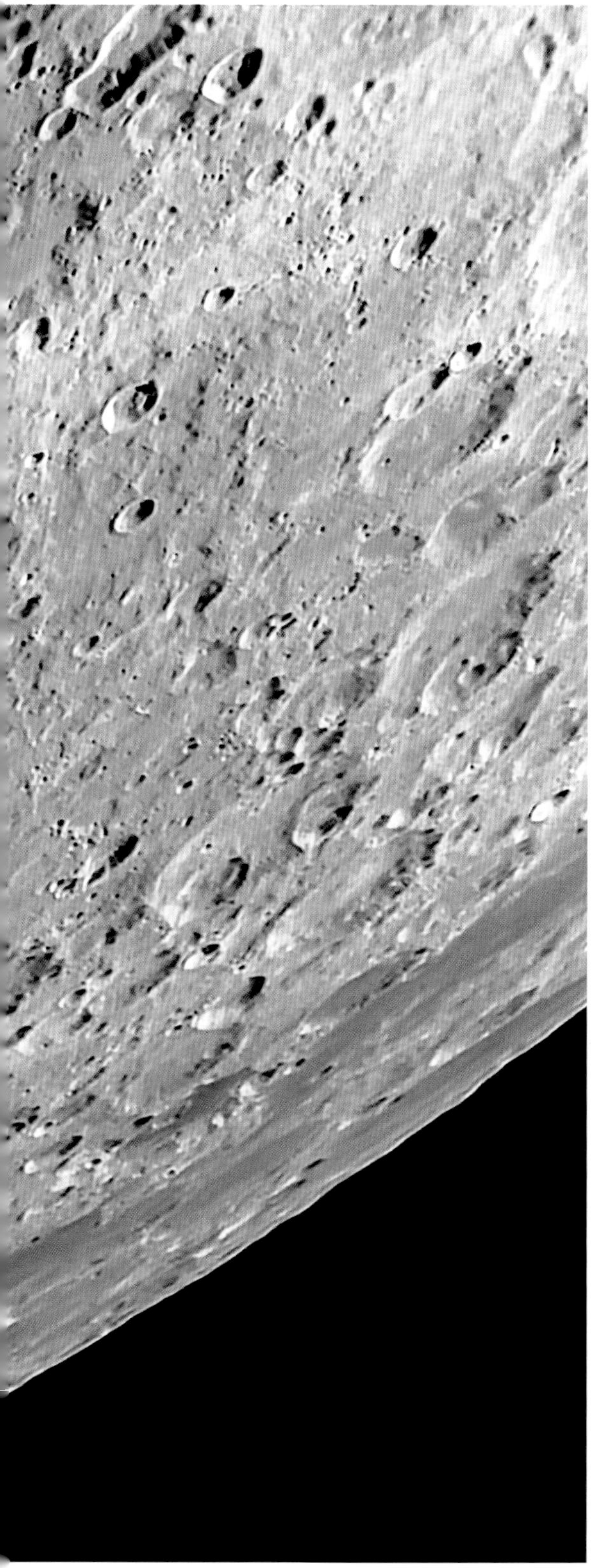

Janssen

At four billion years old, Janssen is the fifth largest crater on the near side of the Moon and among the earliest surviving lunar features. Janssen is a crater that you almost don't notice. Despite its 201-kilometer width, Janssen's floor only dips two kilometers below the surface. Under low sun illumination, shadows make Janssen appear treacherous, but under shadowless high sun, Janssen's huge pie-pan shape becomes featureless and blends into the surrounding highlands. Janssen is heavily ruined but, once recognized, it presents a fascinating contour bisected by the scimitar-shaped 120-kilometer-long rilles of Rima Janssen and the 78-kilometer-wide crater Fabricius. Rima Janssen is often spotted before Janssen itself. The craters Brenner and Metius on Janssen's northern rim create the caricature of bunny ears atop Janssen.

The physical nature of Janssen has evolved over the years as we understand more about the geology of the Moon. Originally classed as a crater type called a walled plain, today we recognize Janssen is a double ring impact basin. Under proper lighting conditions, a trained eye can spot faint evidence of the outer impact ring beyond Janssen's traditional wall.

East of Janssen, the gouge of the 500-kilometer-long Rheita Valley points like a finger toward the mottled expanse of Mare Australe on the Moon's southeastern limb. The valley is named after the 70-kilometer-wide crater Rheita, seen overlapping the northeastern reach of the valley. The southern part of the valley makes a dogleg to the west midway down its length. The upper part aligns with Mare Nectaris, the center of which lies about 700 kilometers northwest. It is easy to imagine the valley being gouged 3.9 billion years ago by massive blocks of material thrown from the Nectaris Basin impact. But it is also hard to imagine those flying mountains changing course midway down the valley. Thus, the valley is not one continuous feature. Some observers have suggested a separate name for the dogleg portion: the Mallet Valley, so named for 58-kilometer-wide Mallet crater west of the dogleg's pivot.

Rupes and Rima Cauchy

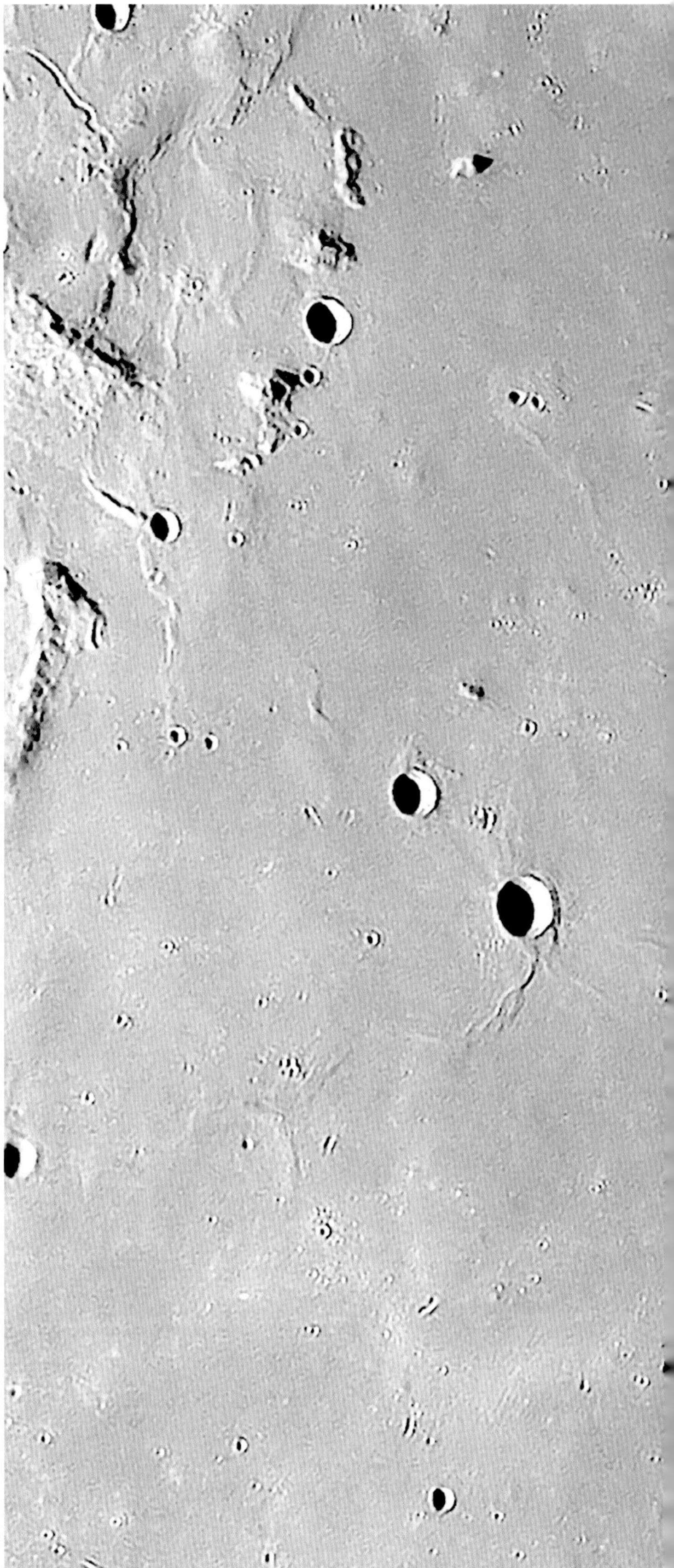

At first glance, the view along the eastern shore of Mare Tranquillitatis looks like a benign area on the Moon, but once the dynamics of the region are understood, it takes on a more exciting personality.

The 13-kilometer-wide crater Cauchy is bracketed north and south by the parallel linear features of Rima and Rupes Cauchy. While each of the long streaks superficially looks similar, each was created by a different volcanic process. The linear rille of Rima Cauchy on the north is a feature called a graben, or an area of land that slumped between parallel faults. While spanning 180 kilometers in length and up to three kilometers wide in places, Rima Cauchy is only about 50 meters deep.

South of Cauchy crater lies the scarp of Rupes Cauchy. Also volcanic in origin, Rupes Cauchy drops 300 meters from north to south and formed when the land collapsed on one side of a single fault line. Rupes Cauchy spans 170 kilometers in length.

South of the twin linear features, the lunar domes, or small shield volcanoes, of Cauchy Tau (west) and Cauchy Omega (east) with a small caldera, rise like small bumps on the smooth plains of Tranquillitatis. These shallow volcanic domes were formed when slow eruptions of very fluid lavas quickly flowed from the eruption point and created low pancake-shaped shield volcanoes.

More evidence of regional volcanism can be seen in the ghost craters dotting the lower right shoreline. Here, the protruding crowns of preexisting craters rise above the basalt that flooded the Tranquillitatis Basin over three billion years ago.

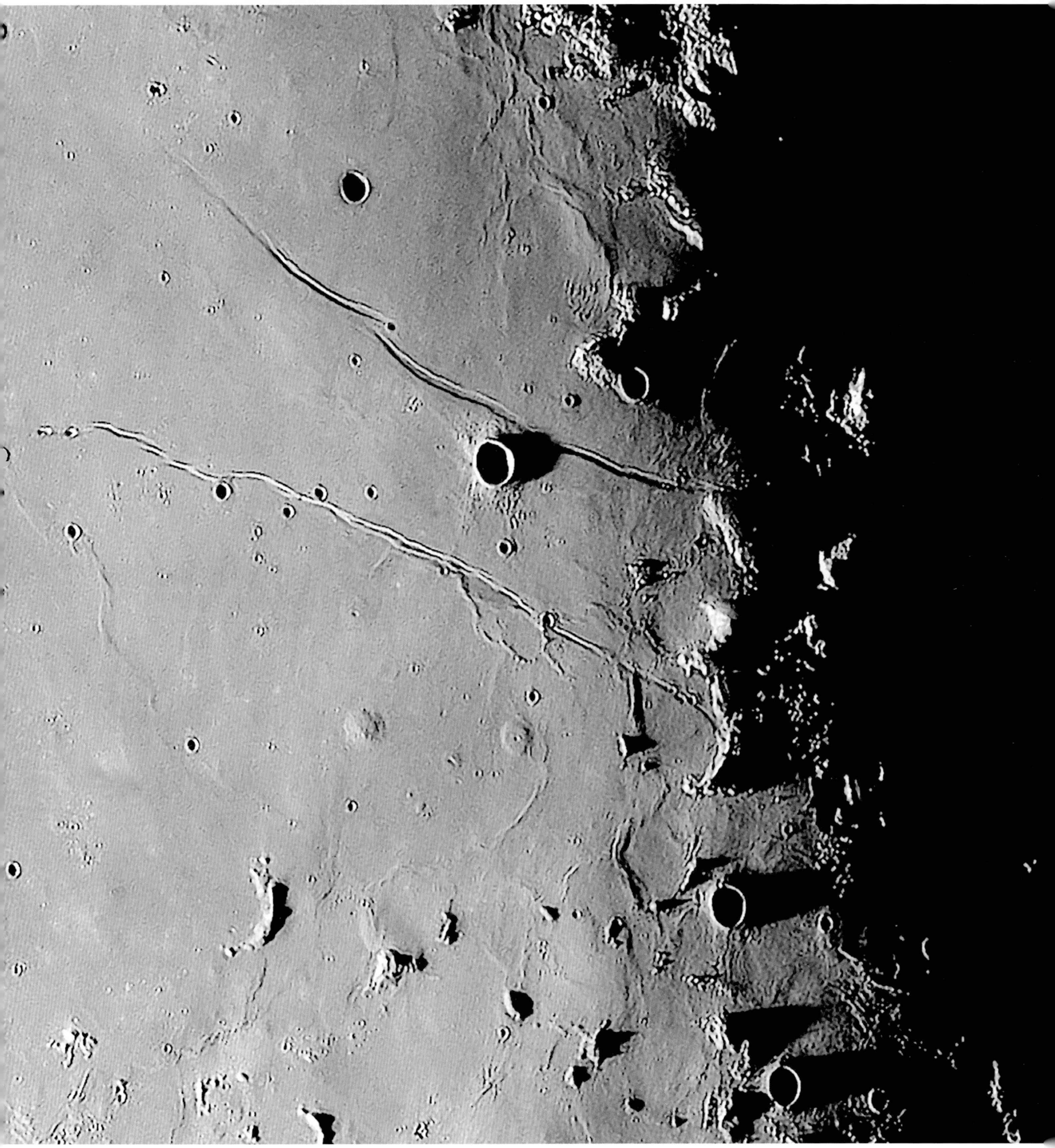

Abulfeda

At image center right, the triple craters Theophilus to the top, Cyrillus and Catharina at the bottom, demonstrate the aging process of lunar features. Theophilus, spanning 101 kilometers, has a sharper and fresher appearance, while 98-kilometer-wide Cyrillus and 101-kilometer-wide Catharina to the south are progressively softer and rounded. Both Cyrillus and Catharina date from about 3.8 to 3.9 billion years ago. Theophilus formed later between one and three billion years ago. The three craters had a significant effect on each other. Catharina was created first, then Cyrillus was blasted into existence and showered Catharina with debris. As much as a billion years later, the explosive formation of Theophilus showered more debris on both Cyrillus and Catharina, further degrading them both.

Theophilus crater overlays the inner Nectaris basin impact ring that forms the shoreline of Mare Nectaris. Cyrillus and Catharina craters progressively arc into the moat between the inner and second Nectarian impact ring. Traces of the second impact ring can be seen extending southeast from Catharina crater. Twenty-eight-kilometer-wide Madler crater lies east of Theophilus. South of Madler lies the strewn field of secondary craters created by the ejecta thrown from the Theophilus impact.

The rough highlands west of Theophilus host the distinctive crater chain Catena Abulfeda extending southeast for 210 kilometers from dark-floored 63-kilometer-wide Abulfeda crater. The lengthy catena streaming from Abulfeda's southern rim inspires the name of "the rat-tailed crater."

North of Abulfeda lies the heavily ruined geologic tangle of 48-kilometer-wide Descartes, a crater that looks more like one of grandma's oatmeal cookies than a traditional crater. The Apollo 16 landing site lies north of the light-colored deposits on the northern side of Descartes.

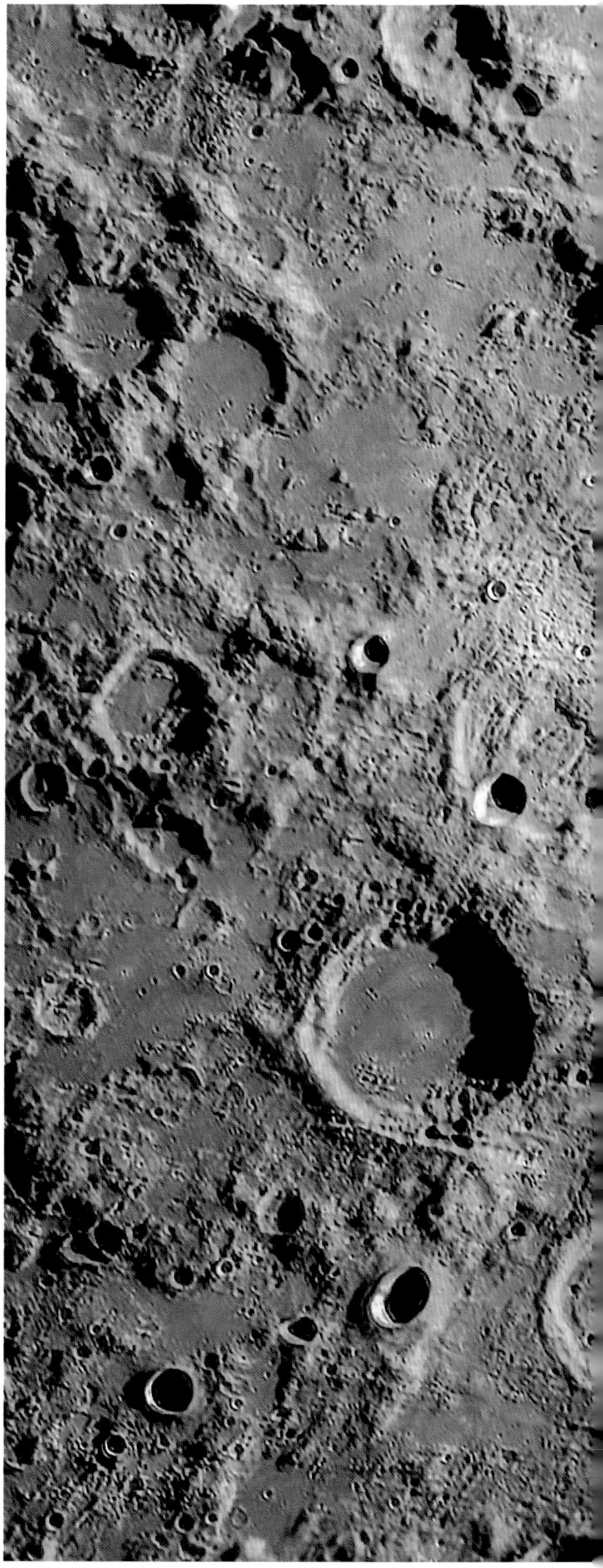

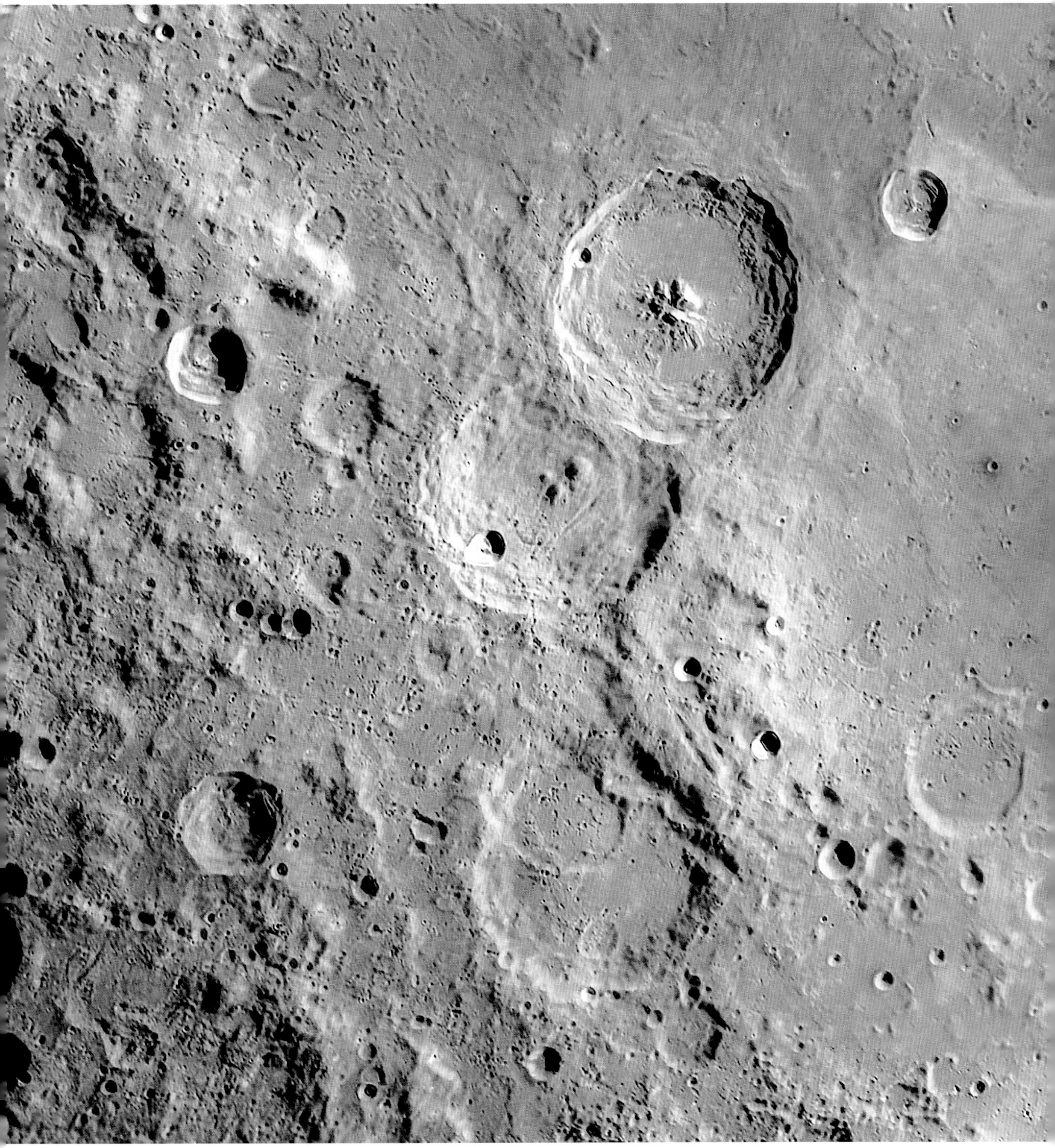

Aristoteles and Eudoxus craters lie at the image-upper-right and span 88 and 68 kilometers, respectively. Although they differ in age by a billion years, Aristoteles and Eudoxus are considered an observational pair. Over time, both have lost their ray structures, but Aristoteles still displays streams of secondary craters and ridges of ejecta across the plains of Mare Frigoris, while similar features around Eudoxus are muted by the highlands north of Mare Serenitatis, seen at the lower right. The interior and central peak within Aristoteles have been covered by impact melt created by the heat of the asteroid impact that formed the crater. The interior of Eudoxus displays a broad expanse of multiple central peaks. The crater pair's proximity to the diverse features on northern Mare Imbrium enhances their observational appeal.

The rugged Alps mountains dominate the upper left with the slash of the Alpine Valley cutting through the elevated terrain. Initially, when it became suspected that massive asteroid impacts created the basins and craters on the Moon, it was thought the Alpine Valley was ripped open by huge chunks of debris heaved by an asteroid impact. Today we realize the creation of the Alpine Valley was more benign. The feature is a graben, or a geological structure created by the slumping of land between two fault lines. Subsequent lava flooding smoothed the valley's floor.

Cassini crater at the lower-left-center presents a unique bird's nest appearance. Cassini formed on the northeast floor of the Imbrium Basin prior to the lava flooding that created Mare Imbrium. As subsequent lava flows smoothed the floor of the Imbrium Basin, the lava lapped to the base of Cassini's outer wall, obliterating the crater's rays and radial streaks of secondary craters. Subsequent lava eruptions through fractures in Cassini's floor flooded and smoothed the interior of the crater. The later impacts that formed the satellite craters Cassini A and B created the illusion of two eggs in a bird's nest.

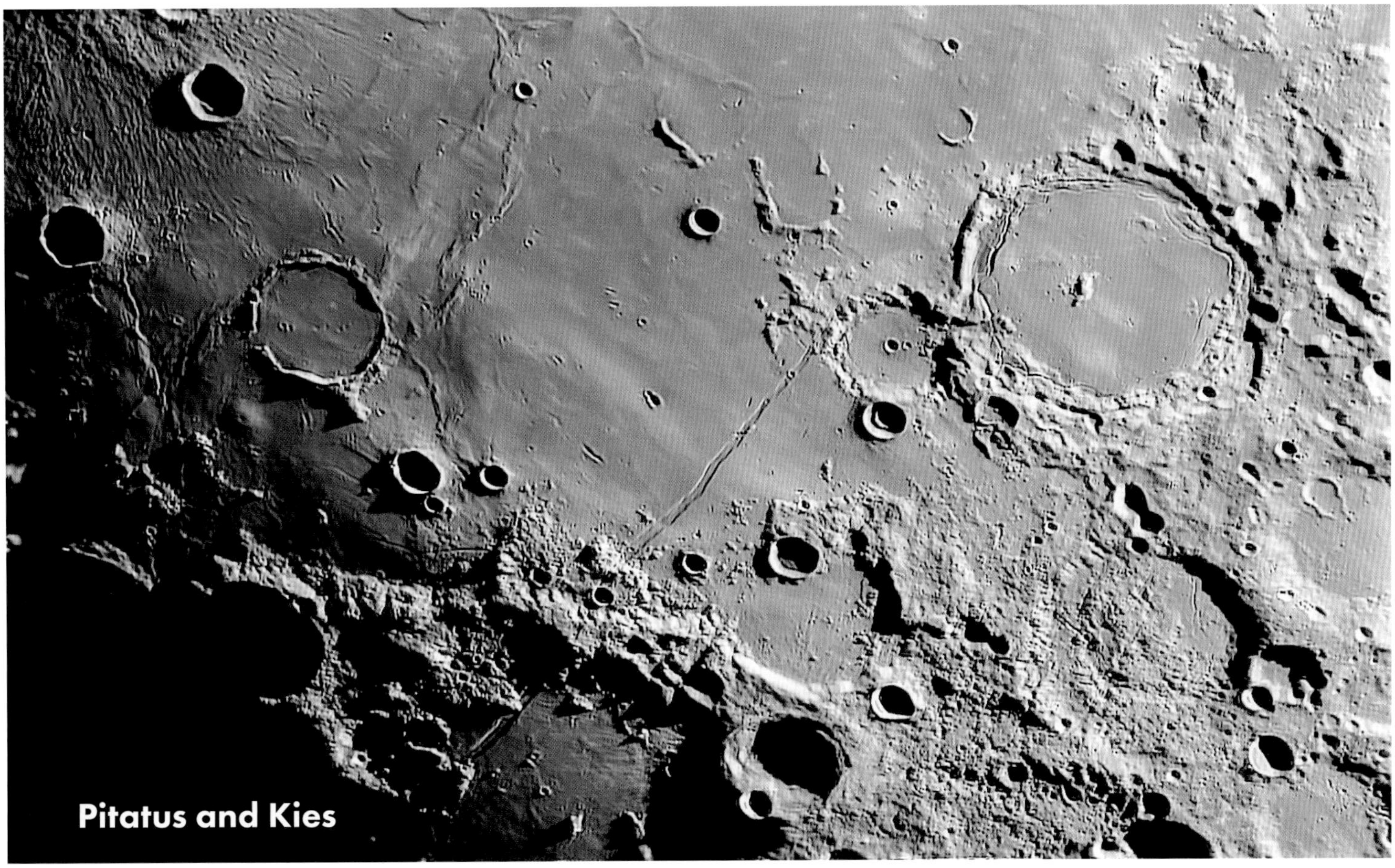

The 98-kilometer floor-fractured crater Pitatus dominates the southern shore of Mare Nubium. Volcanic eruptions through fractures in Pitatus' floor elevated the basalt field within the crater to 150 meters above the nearby plains of Mare Nubium. The volcanic uplift cracked the basalt field within the crater, forming the lengthy circular rilles of Rimae Pitatus, and obscuring all but the upper half-kilometer of the central peak. Two dusty white streaks splashed across the interior of Pitatus are ray material thrown from Tycho crater 350 kilometers to the southeast. The Italian Jesuit monk Giovanni Ricciolli named this feature in 1651 after his fellow countryman and mathematician Pietro Pitati.

Smaller 43-kilometer-wide Hesiodus crater butts against Pitatus' western wall and displays the bullseye impact of five-kilometer-wide Hesiodus D. The 13-kilometer-wide double-walled concentric crater Hesiodus A is nested on the southern rim of Hesiodus.

The 300-kilometer-long slash of Rima Hesiodus marches across the highlands separating Mare Nubium from Palus Epidemiarum to the south. The feature is a graben, or the slumped land between two fault lines.

A curiosity is the wrecked remains of the crater Wurzelbauer south of Pitatus. This 88-kilometer-wide feature is four billion years old and displays a devastated interior that suggests a smaller crater was nested within it before subsequent impact destruction befell the region.

Many ghost craters dot the southern Nubium shoreline. Ghost craters are preexisting craters that were subsequently buried by the lava flows that created the basalt plains of the lunar maria. The crown of the ghost crater Kies is an unmistakable 44-kilometer ring on the southwestern Nubium shoreline. Southwest of Kies lies the raised bump of the shadowed 14-kilometer-wide and 150-meter-high volcanic dome popularly known as Kies Pi.

Hyginus and Ariadeaus

A view of the center of the Moon's visible disk reveals multiple geologic features that add spice to the scene. Three rille systems cross the center of the image: the K-shaped Rimae Triesnecker system lies at image-lower-left, the gull-wing shape of Rima Hyginus hinges on 10-kilometer-wide Hyginus crater, and the linear form of Rima Ariadaeus extends toward Mare Tranquillitatis at the image right.

The smeared appearance of the dark hilly region, the footprint-shaped craters Boscovich and Boscovich P, and circular 91-kilometer-wide Julius Caesar crater at the upper left are due to the avalanche of ground surge debris pushed southeast from the Imbrium Basin impact about 3.8 billion years ago. South of Rima Ariadaeus, the 46-kilometer-wide bullet-shaped crater Agrippa and the 35-kilometer-wide bell-shaped crater Godin demonstrate that not all craters are round.

The southwestern shore of Mare Tranquillitatis hosts the spider-like feature composed of wrinkle ridges known as Lamont. Initially classed as a ghost crater, or a crater that was subsequently buried under lava flows, the 75-kilometer-wide circular feature is now regarded as a ghost basin due to the faint outer ring surrounding Lamont's eastern side. The heavily shadowed 27-kilometer-wide crater Arago lies west of Lamont and highlights the nearby mounds of the shallow volcanic Arago Domes. The Apollo 11 landing site lies between Lamont and the twin fractures of Rimae Hypatia along the Tranquillitatis shoreline.

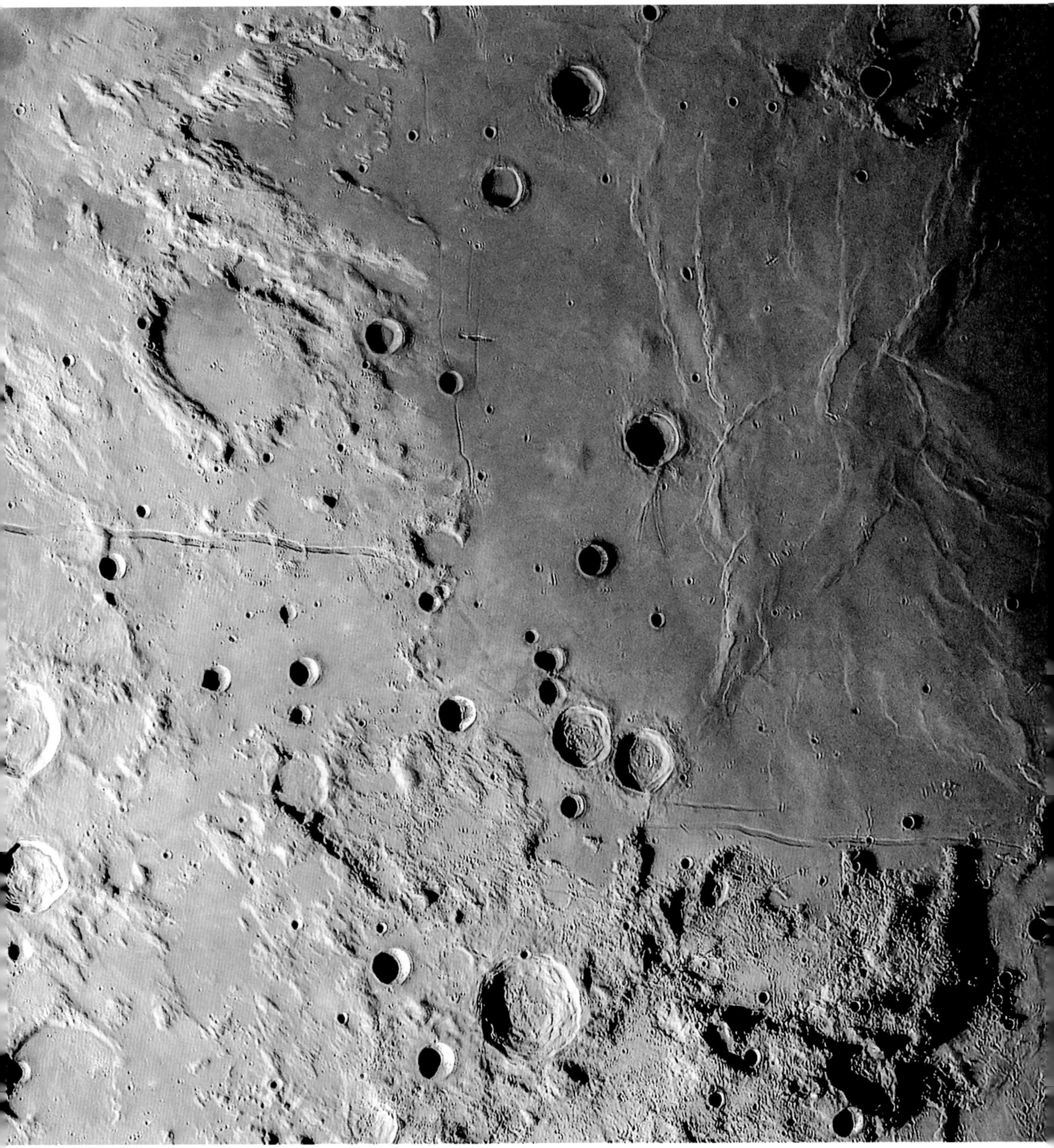

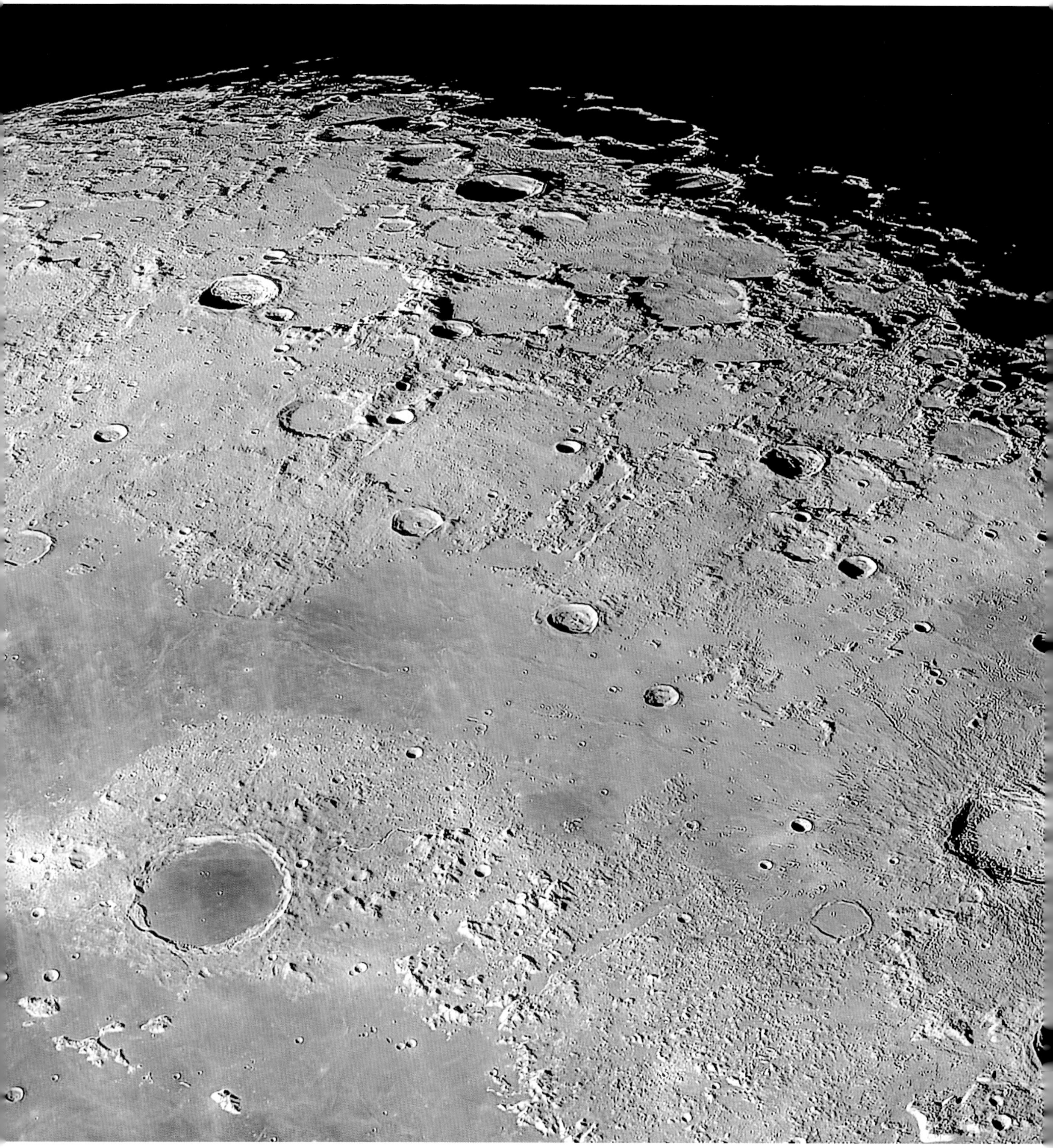

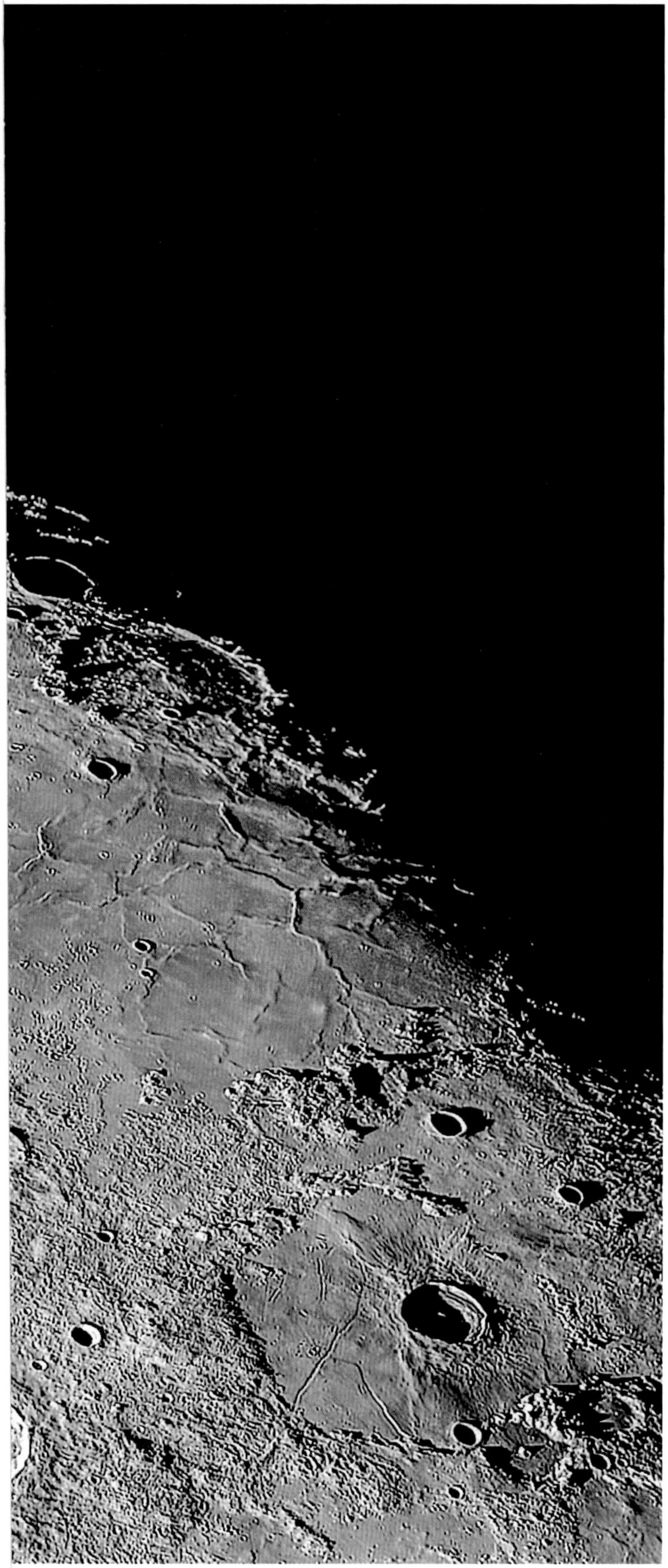

Frigoris East

The polar territories north of Mare Frigoris, the Sea of Cold, are rarely explored by the casual observer, but many observational treats await those who visually trek into the region. The northern territories are unofficially bordered by Mare Frigoris, the ribbon-like mare that spans 1,800 kilometers in length but is only 200 kilometers wide and forms the eyebrows above the Man-in-the-Moon's eyes.

Eastern Mare Frigoris is relatively bland and smooth. As it arcs above the dark flat floor of Plato crater and meanders eastward, the only significant crater to mar its surface is 88-kilometer-wide Aristoteles on the southern shoreline. Pronounced wrinkle ridges crisscross eastern Mare Frigoris. Wrinkle ridges typically form where sheets of mare basalt are crushed together as they slump toward the depressed center of a basin. The presence of wrinkle ridges within Frigoris indicates the basalt within the narrow mare may be deeper than its narrow size would suggest.

Northeast of Plato, across Mare Frigoris, the square shape of 159-kilometer-wide W. Bond crater hugs the northern shoreline and presents the form of fortress-like ramparts. Continuing northeast, the overlapping craters Meton and Meton C, D, and E combine to form a cloverleaf pattern collectively known as Meton. West of Meton, the ancient territory has been ruined by the infall of ejecta blasted out of the Imbrium and Serenitatis basins south of Mare Frigoris. Most craters in the rough-and-tumble region are smooth-floored, debris-filled shallow features punctuated by the crisp appearance of 51-kilometer-wide Anaxagoras, the only crater in the region that is less than a billion years old.

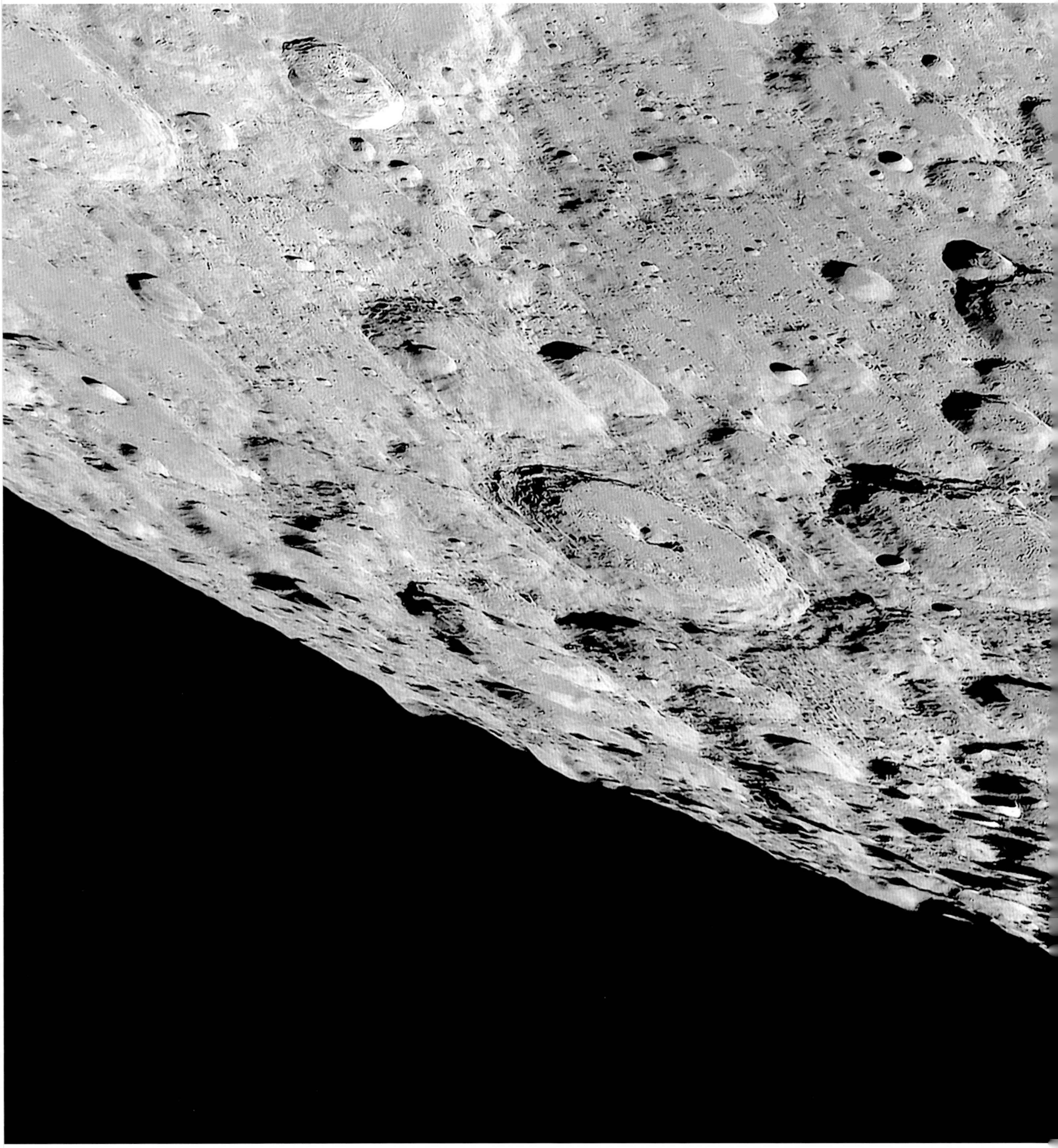

Moretus

The field of ancient ruined craters on the southern highlands is marked by the unusual fresh appearance of Moretus crater, seen near image center. Spanning 114 kilometers, Moretus may be as old as three billion years, but despite its age, Moretus was among the last of the large impacts in the region. Moretus thus stands out for the clarity of its geologic form. The preserved one-kilometer-high crater rim rises above the surrounding territory, then descends across collapsed terraced walls to the flat floor lying three kilometers, which is centrally punctuated by a two-kilometer-high central peak.

The resemblance between Moretus and the landmark crater Tycho is superficial because Moretus lacks Tycho's amazing hemisphere-wide ray system. The two craters share nearly identical size and form, but Moretus is much older than Tycho. Crater rays fade after about a billion years, so Moretus now stands alone on the heavily cratered southern highlands. The crater gained its name in 1651 when Riccioli named the feature after Theodore Moret, a 17th-century Belgian mathematician.

Our oblique Earthly view of Moretus allows us to view the far southern terraced crater wall almost as if we were standing on the crater's opposite rim and enjoying the geologic spectacle. Equally striking are the mountain peaks projecting from the southern limb against the blackness of space. Some of these lumps and bumps are portions of crater rims while others are part of the outer rim of the South Pole–Aitken Basin which spans from the equator to the South Pole on the back side of the Moon. The massive formation is the largest known impact basin in the solar system. Fleeting glimpses of its rim can be seen along the southern limb, with some peaks rising six kilometers.

Rupes Recta

The unique linear feature on the western shore of Mare Nubium is affectionately called Straight Wall and is the most beloved non-crater formation on the Moon. Officially designated Rupes Recta, Latin for "straight cliff," the feature appears as a black line when it casts a sunrise shadow, then as a white line when its slopes are illuminated at sunset.

Depicted as an imposing cliff prior to the Space Age, Rupes Recta is a steep slope, not exceeding 30 degrees, descending a maximum of 350 meters from east to west. The slopes were formed by the collapse of the surface along the western side of a fault line. The feature spans 116 kilometers in length but rarely reaches 800 meters in width along its slope. The southern end is intercepted by a hilly region and the arc of a partially buried crater rim. The two formations allow the mind's eye to see Straight Wall as a sword with the crater rim forming the handle and the hilly outcrop forming the handguard. Collectively, the formations are often called "Huygens' Sword" because the feature was first studied by Christiaan Huygens in the 17th century.

Straight Wall lies within a large unnamed crater whose eastern wall arcs around Straight Wall to form a horseshoe bay on the shore of Mare Nubium. Straight Wall's host crater lies on the tilted rim of the Nubium Basin, thus the lower western side of this anonymous crater has been submerged by lavas that flooded Mare Nubium over three billion years ago. Evidence of the hidden rim is seen by the curved wrinkle ridges that mark the crater's full circular form. Spanning nearly 200 kilometers, this half-hidden unnamed crater is one of the Moon's larger craters and begs for a name of its own. Unofficially it is designated Ancient Thebit, a name derived from 58-kilometer-wide Thebit crater which overlies the raised rim of the horseshoe bay.

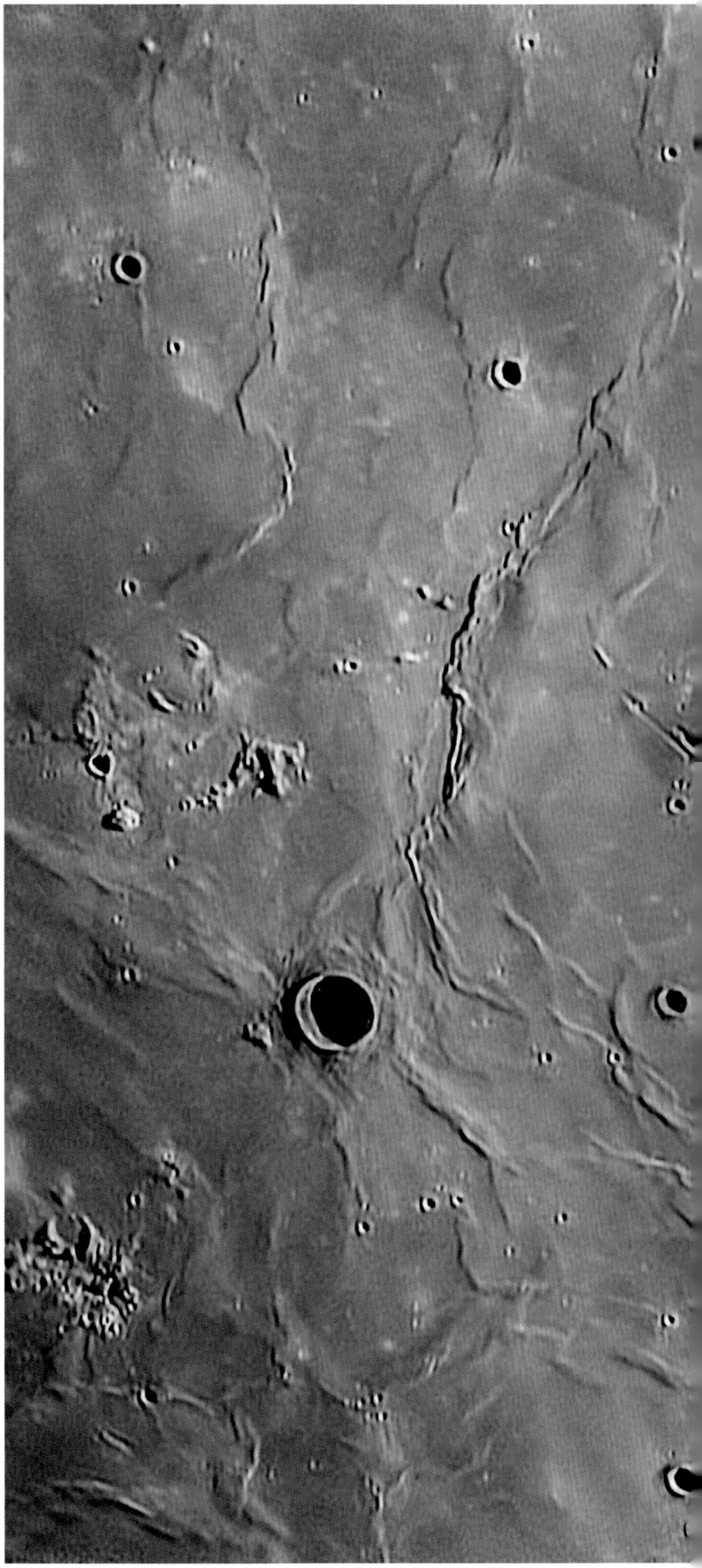

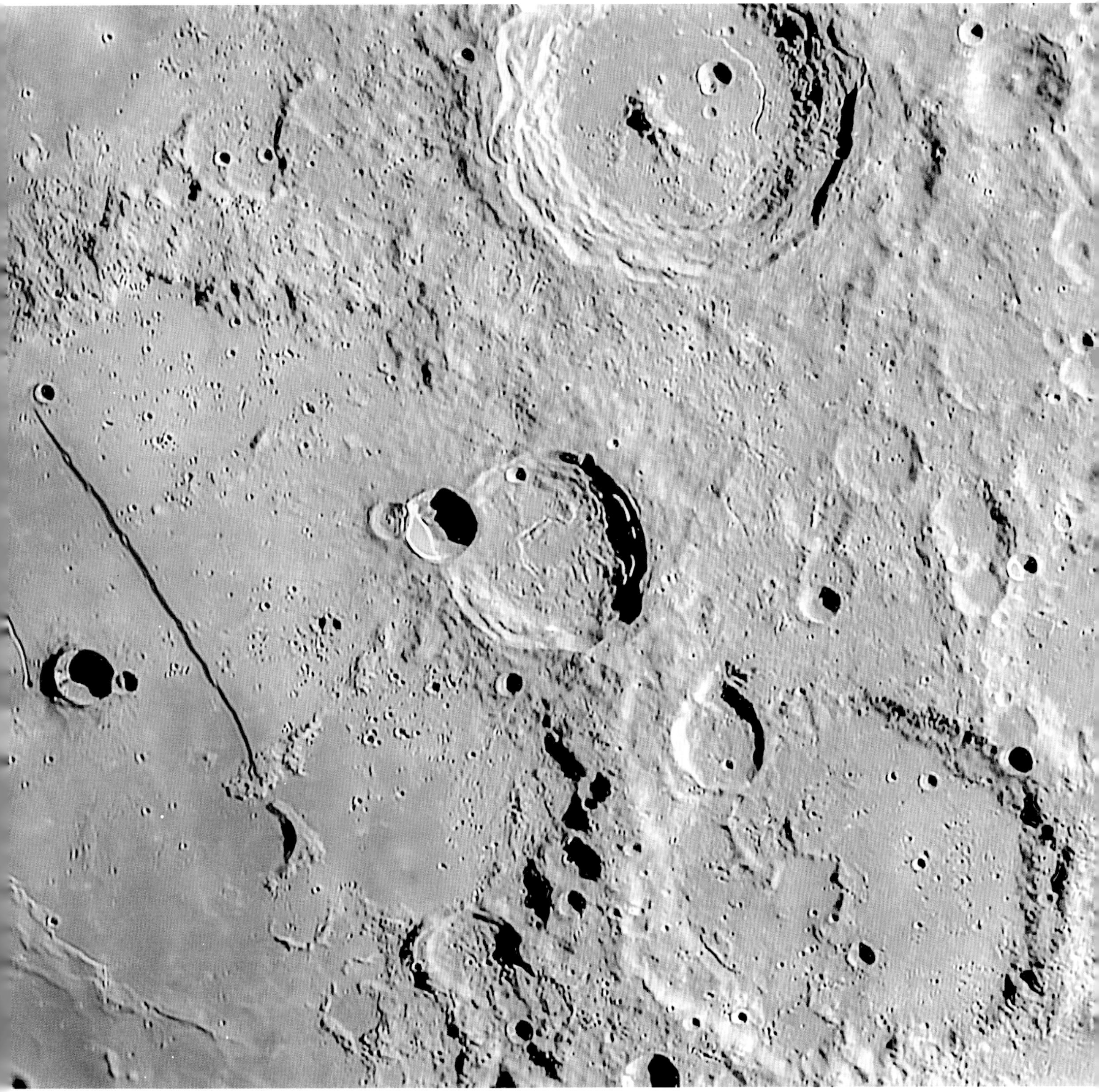

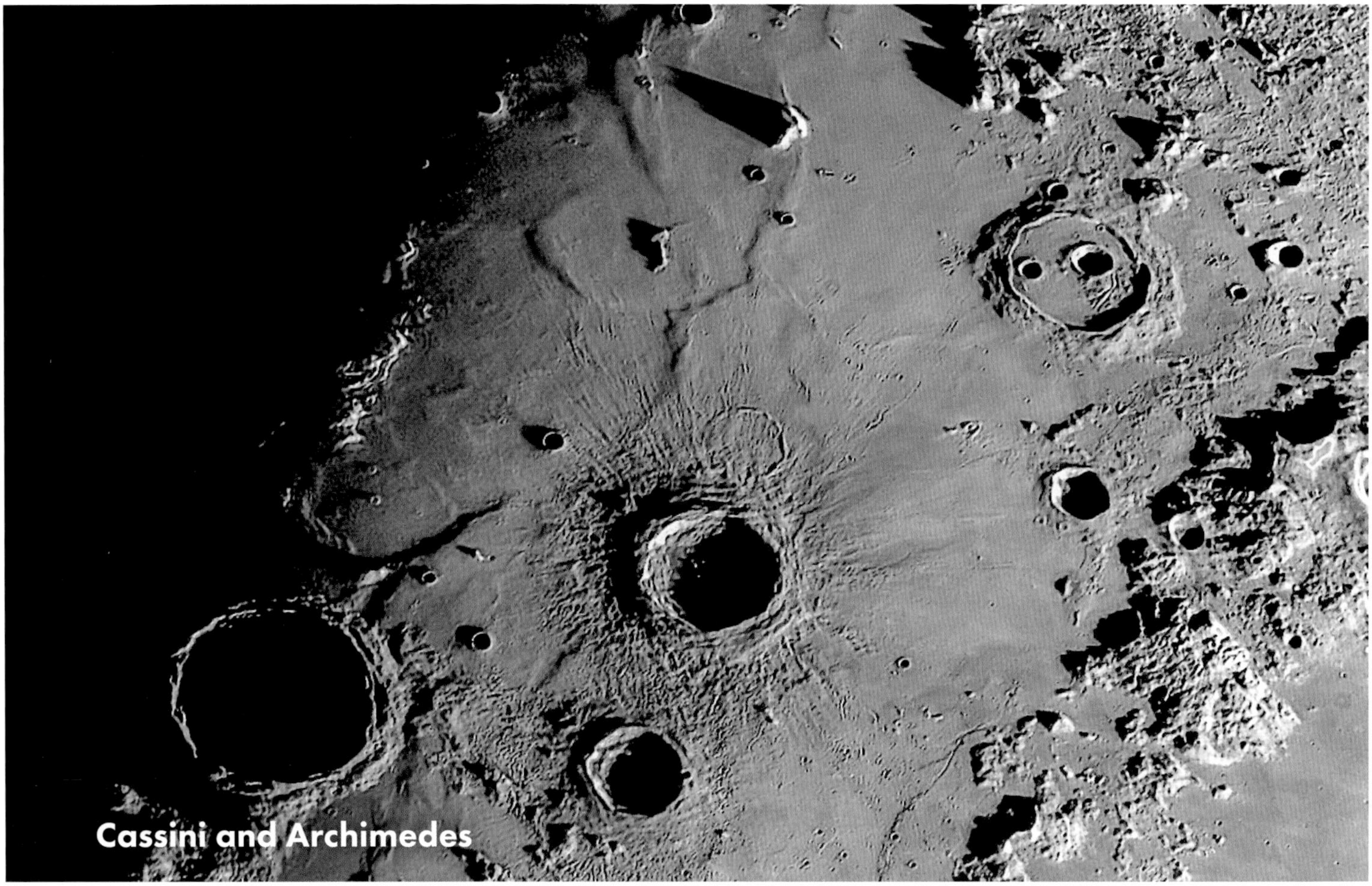

The bird's-nest form of Cassini crater anchors the view at upper right, but the triple crater formation near the terminator grabs our attention. The symmetrical rim of 83-kilometer-wide Archimedes crater protrudes into the sunrise and joins equally shadowed 55-kilometer-wide Aristillus and smaller 40-kilometer-wide Autolycus to create the triangular formation. Curiously, all three craters have names that begin with the letter A.

In the short history of Earth being a spacefaring world, this region of the Moon holds historical significance. In September 1959, the triangular crater formation was the target where the first spacecraft from Earth reached the Moon when Luna 2 impacted between Archimedes, Aristillus, and Autolycus. To honor the technical feat of being first to the Moon, the region between the craters was renamed Sinus Lunicus, or the Bay of Lunik, as the spacecraft was called in the press.

Shadowed Archimedes crater, seen at lower left, is the largest crater within the expanse of Mare Imbrium. Sunrise illumination on Aristillus crater accentuates the broad fan of washboard ground surge pushed outward by its creation about a billion years ago. A thin finger of secondary craters reaches from Aristillus to Cassini crater at the upper right. A nameless, partially buried 30-kilometer-wide ghost crater lies on the north face of Aristillus and has curiously escaped being officially named. Autolycus crater is similar in age. Sawtooth shadows from the lower reach of the Alps Mountains at the upper right and the Apennines jutting in from the lower right illustrate the ruggedness of the mountain chains. The lonely peak Mons Piton at the top center casts a long shadow. Midway between Mons Piton and Aristillus lies the T-shape of Thor's Hammer, so nicknamed for its resemblance to the magic hammer wielded by the Norse god Thor.

Plato and the Alpine Valley

The rugged Alps Mountains extend east of 101-kilometer-wide Plato crater. The startlingly smooth interior of Plato crater rivets the attention simply by its sheer lack of detail compared to other regions of the Moon. Contrasting with the complexity of the Alps Mountains and the gash of the Alpine Valley bisecting the Alps, the black-floored crater stands as a mesmerizing blank anomaly on the scarred face of the Moon. A 24-kilometer-long triangular block of material fractured away from Plato's western wall but remains wedged in place by the basalt fill within Plato. Curiously, the flat expanse of basalt within Plato stands 600 meters higher than the nearby plains of Mare Imbrium.

The Alpine Valley is one of the anchor features on the Moon's face that both novice and experienced observers enjoy. The valley is a geologic feature called a graben, or land that has slumped between two parallel faults. Stretching 155 kilometers and spanning 15 kilometers in width, the floor of the Alpine Valley descends 800 meters from the surrounding hilly terrain. A fascinating 500-meter-wide sinuous rille runs the length of the Alpine Valley. The channel was carved by flowing lava during the Moon's volcanic period over two billion years ago. Curiously, despite the rille's prominent location within the Alpine Valley, it remains unnamed.

Progressing inland into the Alpine Valley from the plains of Mare Imbrium, the mouth of the unnamed bay at the western end of the valley funnels down from 21 kilometers to a gorge 200 meters wide at the entrance to the valley. The narrow channel is bound by a northern massif that rises 1.8 kilometers, while the southern massif rises three kilometers above the plains of Mare Imbrium. Both massifs are reminiscent of massive ancient Egyptian monuments guarding the path to the Pharaoh's paradise.

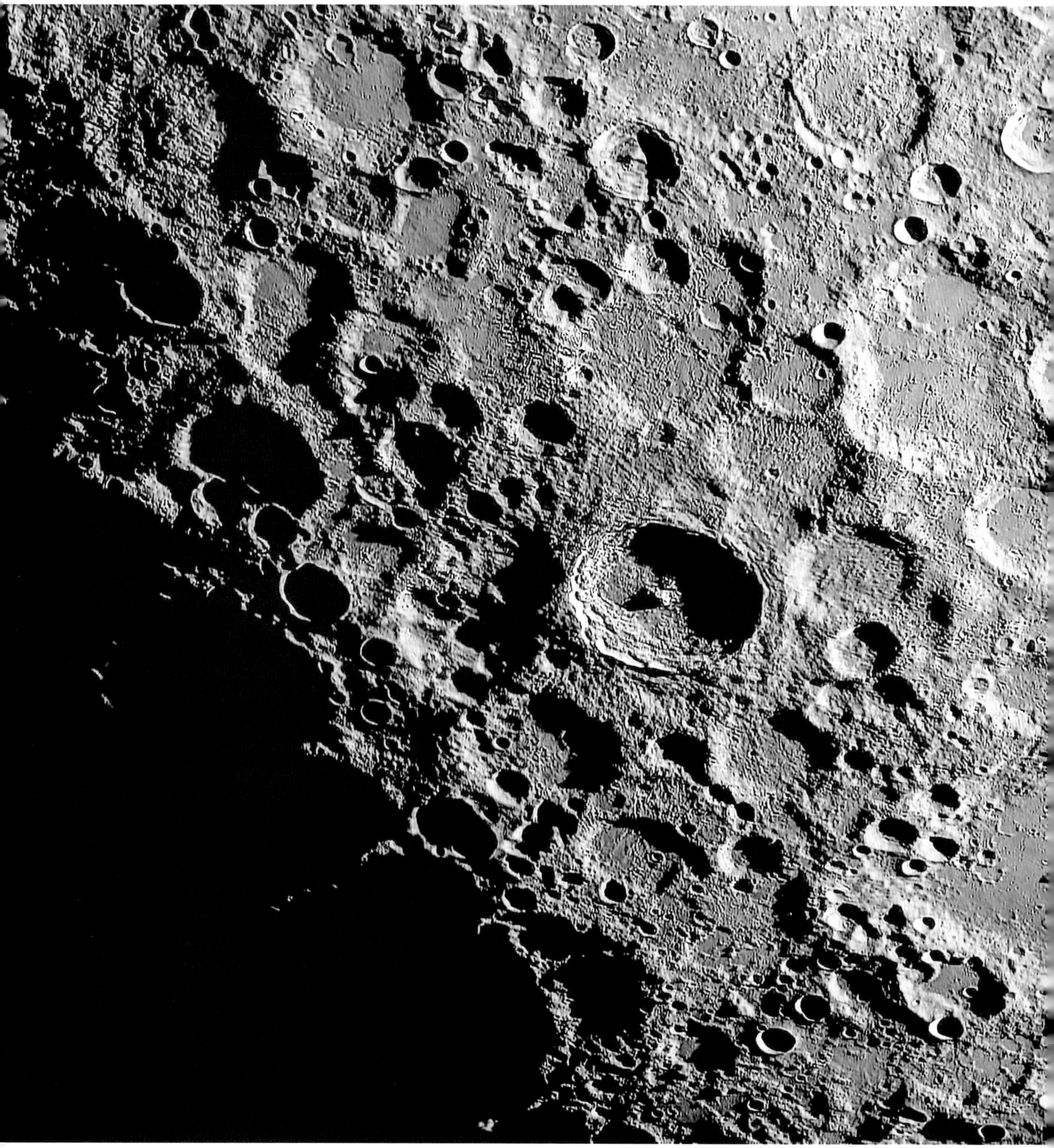

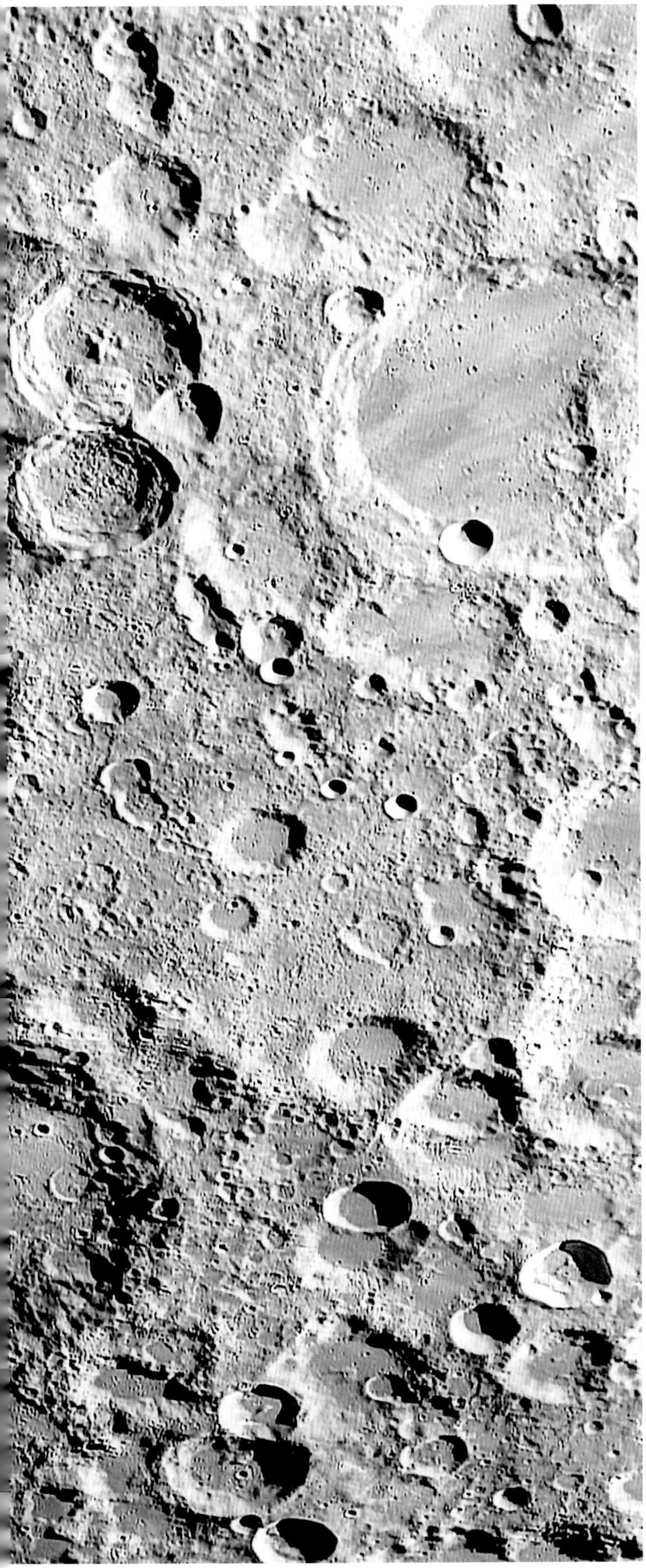

Tycho secondaries

The impact that created Tycho crater 108 million years ago carved the young crater out of territory crammed shoulder to shoulder with ancient craters that are as much as four billion years old. The flash of its explosive birth likely startled many a T-Rex here on Earth. The rough surrounding highlands contrast with the sharp and fresh appearance of Tycho.

Tycho crater is noted for its amazing hemisphere-wide ray system, but crater rays are all but invisible at low sun elevation. However, the low sun elevation provides detail-enhancing shadows that allow us to see the hundreds of tiny secondary crater pits in the highlands surrounding Tycho. These tiny craterlets were created by the skyscraper-sized blocks of debris blasted out of Tycho by the impact of an asteroid just 100 million years ago. The lack of subsequent impacts leaves these new secondary craters undisturbed.

The density of the small Tycho secondary craters diminishes with greater distance from Tycho. Many are aligned in linear chains radiating outward from Tycho. The heavier blocks fell closer to Tycho, the parent crater, while the lighter, more powdery debris was flung farther across the surface forming Tycho's famous brilliant ray structure. Curiously, some of the streams of small secondary craters do not align with the middle of Tycho but instead appear tangential to Tycho's rim.

The Tycho-facing rims of preexisting heavily ruined craters east of Tycho are noticeably smoother and more muted than the rims opposite Tycho. The seismic shock of the cosmic thunderclap that created Tycho violently shook the region and features closer to Tycho collapsed in landslides from the rolling seismic wave, while those farther away were spared by the diminishing intensity of the shock wave.

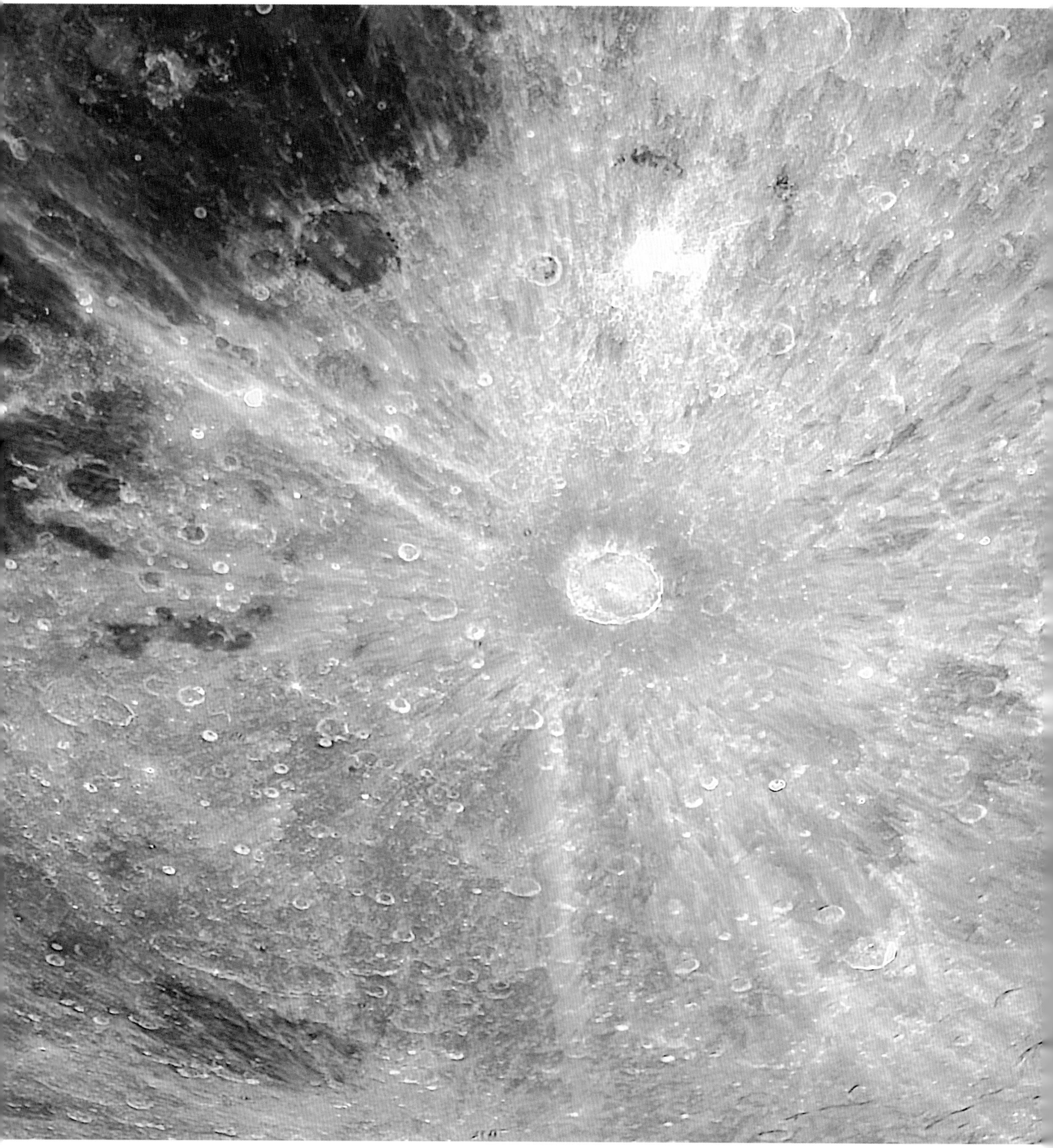

Tycho Ray System

The Tycho crater ray system is the biggest on the Moon and dominates the southern hemisphere. Spanning 2,200 kilometers, Tycho's rays create a naked-eye phenomenon that looks like the navel on an orange. The majority of Tycho's rays spray outward from the middle of Tycho, as we would expect from a feature created by an asteroid impact. But a puzzle is the parallel "railroad tracks" projecting to the northwest that are tangential to Tycho's rim, and the rogue ray extending south that does not align with Tycho.

Crater rays are most easily seen at full moon. The rays are made of crushed, pulverized rock and impact-melted glassy particles blasted out of the crater during its explosive formation. The ray material is like the glassy beaded coating that makes highway signs so reflective in car headlights. The glassy particles are very good at reflecting light back toward its source, your car headlights. The glassy lunar ray material particles similarly directionally reflect sunlight. At full moon, the Earth lies between the Moon and Sun, thus sunlight reflected by the ray material appears brightest during this time.

The inverse of Tycho's bright rays is the dark donut extending about a Tycho radius from the crater's rim. The donut is also most prominent at full moon. The donut is the same gray shade as the relatively ray-free highlands west of Tycho. The donut appears because the territory beyond the donut is brightened by drapes of ray material contrasting with the darker territory within the donut. Although the territory surrounding Tycho was violently shaken by the seismic shock of Tycho's formation, the storm of ejected ray material arced over the donut region, preserving its normal darker color.

Copernicus and Kepler

Sunset shadows emphasize the terraced walls of 93-kilometer-wide Copernicus crater at the upper right. The shallow illumination reveals the folds of ground surge and strings of secondary craters radiating outward from Copernicus. At the left, the impact of 31-kilometer-wide Kepler crater created a 500-kilometer-wide ray system that is the second youngest on the Moon. Kepler's spectacular rays are noted for their unusually thick western spokes and nearly solid fan of eastern rays. The circular splash of rays inspires the nickname of the "Ninja Star." The observationally pleasing rays interact with the Copernicus ray system to create a bright blotch on the Moon that is visible with the naked eye at full moon. The partially shadowed forms of 48-kilometer-wide Reinhold crater and 40-kilometer-wide Lansberg crater lie southwest of Copernicus crater.

The low sunset illumination also demonstrates why the region between Copernicus and Kepler was renamed Mare Insularum, the Sea of Islands, in 1976. Dozens of peaks from the southern slopes of the Carpathian Mountains protrude above the basalt that floods the area, giving the illusion of islands in a vast ocean. Previously the area was part of Oceanus Procellarum, and the designation change shows the Moon's nomenclature is far from static and unchanging.

Sunset illumination enhances the relief of the hills of Mare Insularum west of Copernicus crater and reveals groups of shallow volcanic domes. Four domes, each topped by a small caldera, lie between 15-kilometer-wide Hortensius crater and two of the largest hills rising southwest of Copernicus. Another prominent dome, formerly known as Milichius Pi, lies northwest of Milichius crater. Additional shallow domes, some with calderas, dot the region southwest of Tobias Mayer crater.

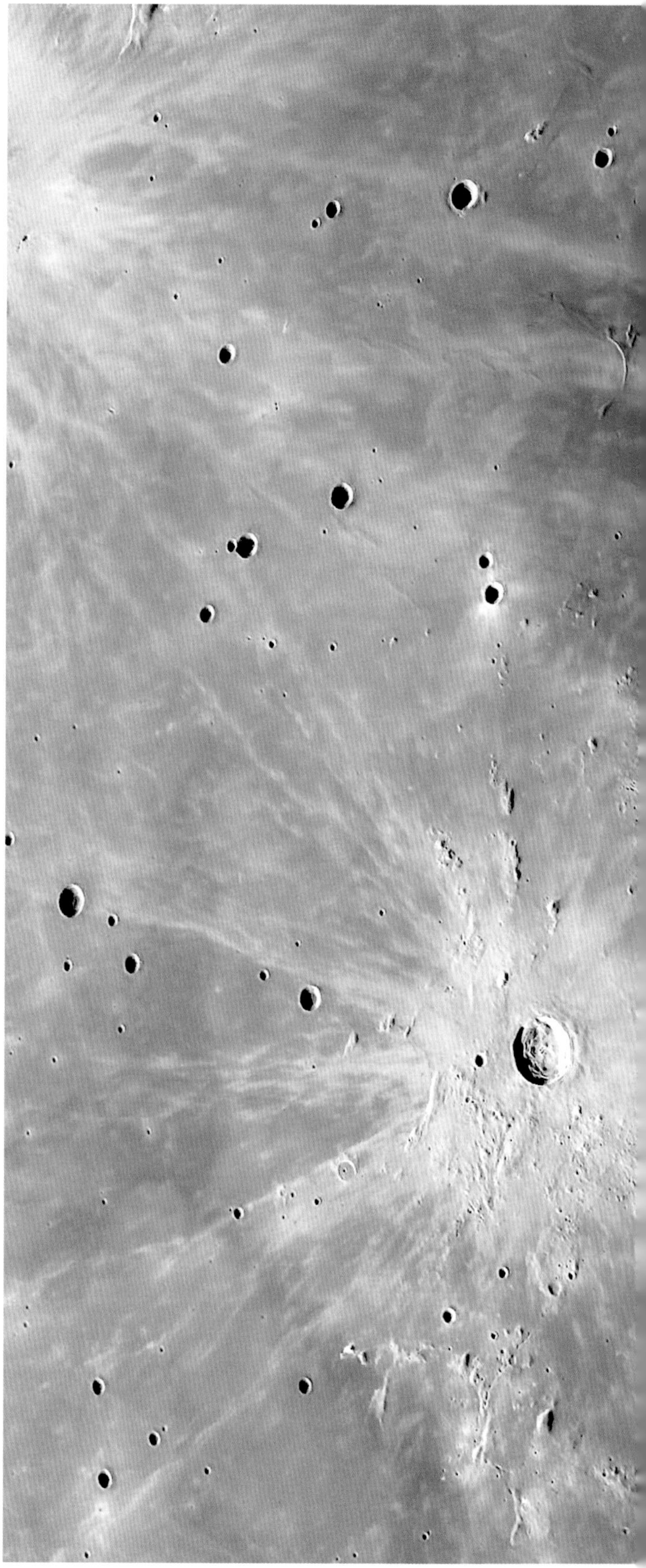

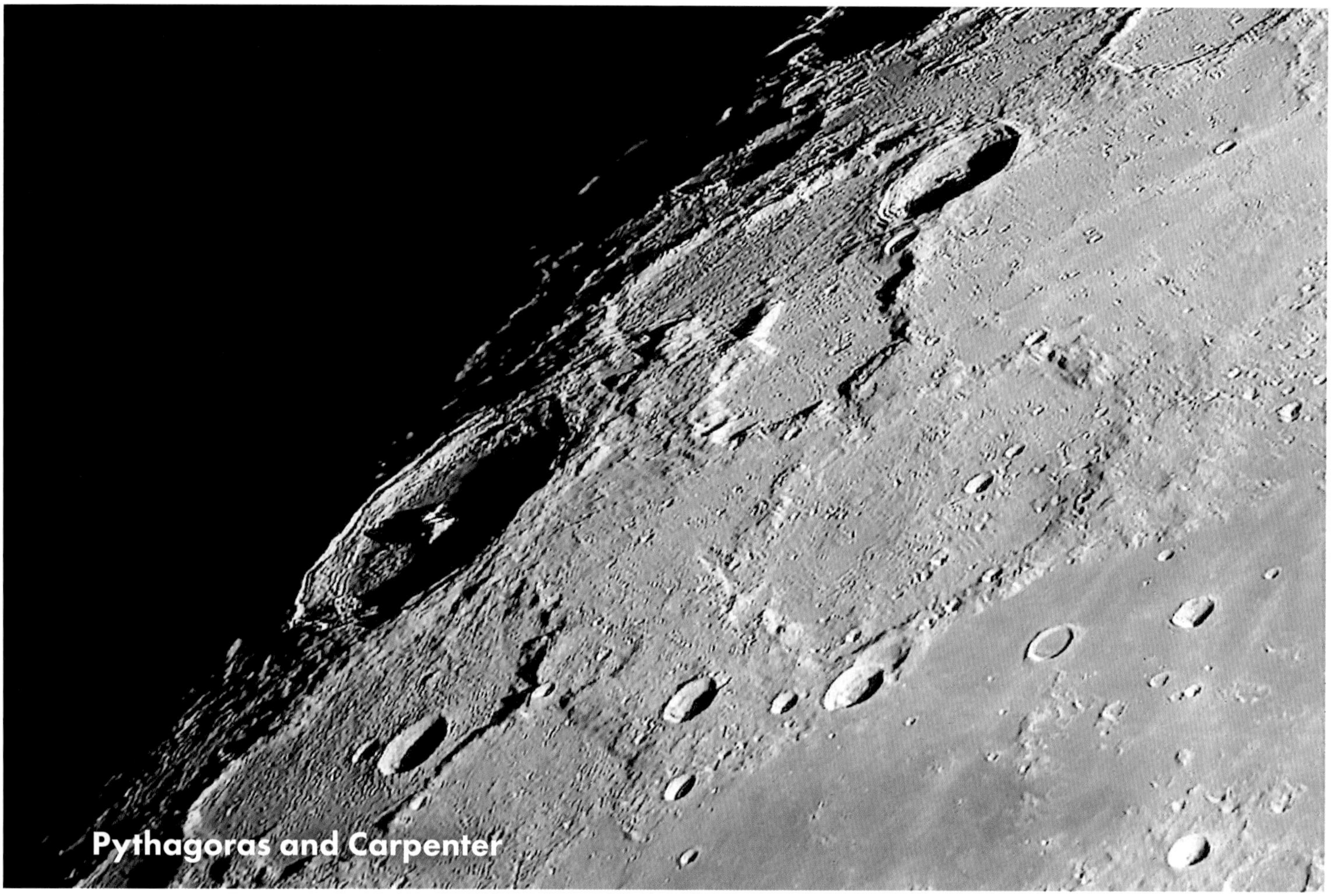

Pythagoras crater emerges from the lunar night shortly before the full moon phase. Pythagoras lies on the northwestern limb of the Moon, and our Earthly view is always at an oblique angle, emphasizing its collapsed terraced walls. Spanning 129 kilometers and nested south of the Valentine heart formed by the merged halves of Anaximander crater, Pythagoras is easy to spot because of its sharp, angular form. Pythagoras is similar in form to the landmark Copernicus crater but is a quarter larger. Pythagoras was created about three billion years ago, after the early collision carnage that preceded Pythagoras' formation, thus it has not been worn and eroded like nearby older craters.

At sunrise, the fascinating prominent central peaks within Pythagoras present a double pyramid shape to our oblique view, but an overhead view shows they are two conventional circular peaks spanning 34 kilometers across their base. The nearside peak rises three kilometers above the floor of Pythagoras and reaches slightly above the plains surrounding Pythagoras. The rear peak only rises 1.5 kilometers above the floor of Pythagoras, which itself lies 2.5 kilometers below the surrounding territory. This makes the twin central peaks statistically a little larger than average and shows the collision that formed Pythagoras was severe, resulting in higher-than-normal rebound in bedrock under the impact site.

The Valentine heart shape of Anaximander B and D craters next to Pythagoras presents an equally fascinating form. Spanning 79 and 96 kilometers, respectively, the outer walls of the debris-filled, merged halves of Anaximander rise nearly two kilometers to create the heart shape anchored on the east by 60-kilometer-wide Carpenter crater. The 156-kilometer-wide ruined form of J. Herschel crater lies in front of Anaximander.

In a sea of edge-to-edge craters, the massive formation of 231-kilometer-wide Clavius crater begs our attention. Located at 60 degrees south latitude on the tortured terrain of the southern highlands, Clavius presents an oblique view of its interior and is unmistakable simply by its sheer size. Clavius was once considered the largest crater on the near side of the Moon, but modern cartography shows Deslandres crater exceeds Clavius' diameter by four kilometers.

In 1651, the Italian Jesuit astronomer Giovanni Ricciolli named Clavius after the 16th-Century German mathematician and astronomer Christoph Klau. Ricciolli must have held Klau in very high esteem to have named what was apparently the Moon's largest crater after him.

When it formed nearly four billion years ago, Clavius likely had one of the largest ray systems to span the face of the Moon, splashing across an entire hemisphere. Before they were erased by volcanism and meteor impacts, the Clavian rays were the crown jewel of the solar system.

Although nearly five kilometers deep, Clavius is so broad that it almost disappears under the high noon sun, but at lower sun elevations it looks dramatically deeper. Dating from nearly four billion years ago, Clavius has been targeted by later impacts that render its form like a work of art created by a talented designer. The western rim of Clavius is punctuated by the small crater Clavius L, which along with Porter on the northeastern rim and Rutherford on the southeastern rim, creates the illusion of a westward-pointing arrowhead. The illusion is strengthened by the artistic arc of Clavius J, N, C and D sweeping eastward across Clavius' floor toward Rutherford crater.

The deeply shadowed pit south of Clavius is the 106-kilometer-wide crater Blancanus.

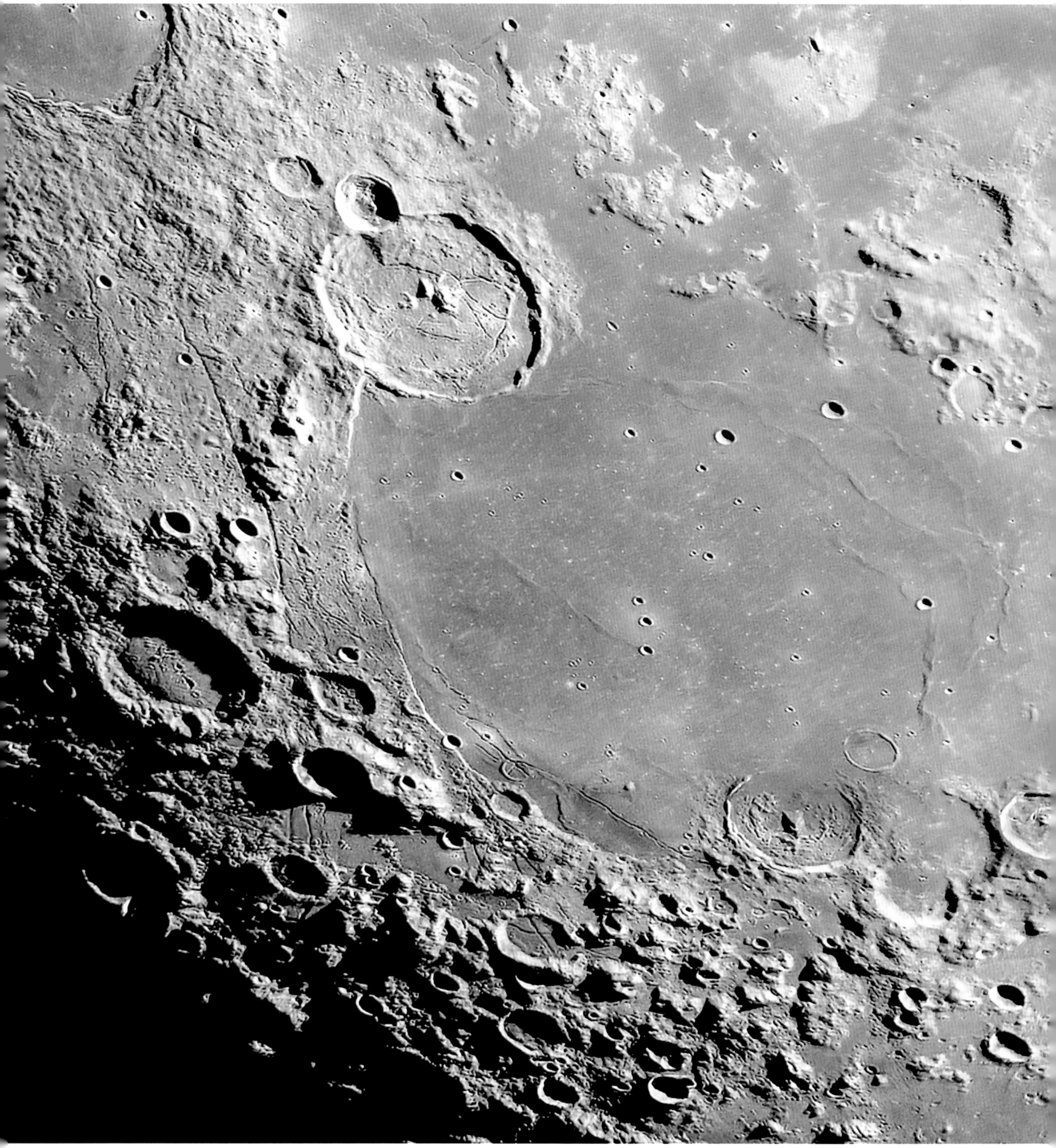

Mare Humorum

The interior of Mare Humorum, the Sea of Moisture, is a bland expanse, but there is much detail in the territory surrounding its dark basalt plain. On the western shore, volcanic dikes pushed up from the Moon's molten core have split huge cracks in the surface. On the eastern shore, the curved rilles of Rimae Hippalus parallel the arc of the shoreline and look like a massive cat scratch across the face of the Moon. Rimae Hippalus was created when the basalt within the Humorum Basin depressed the middle of the basin and stretched the shoreline, pulling open the lengthy curved rilles. On the eastern plains of Humorum, sheets of basalt that were crushed together as they slumped toward the depressed center of the Humorum Basin created a series of unnamed wrinkle ridges that mirror the curve of Rimae Hippalus.

While there are small bright spots of ejecta surrounding small craters within Humorum, this mare is unique in that it is not crossed by a major crater ray streamer. Mare Humorum is one of the Moon's smaller maria and surrounded on three sides by bright highlands, making it easy to see with the naked eye near the full moon. Northeastern Mare Humorum blends into Oceanus Procellarum through a cluttered 150-kilometer-wide channel between the craters Gassendi and heavily ruined Agatharchides.

The northern shore of Mare Humorum is highlighted by the 111-kilometer-wide, floor-fractured crater Gassendi which looks like one of grandma's oatmeal cookies fresh from the oven. On the southern shore, lava-flooded, floor-fractured craters Doppelmayer, spanning 65 kilometers, and Vitello, spanning 41 kilometers, artistically balance Humorum. Floor-fractured craters form when magma pushes upward under the crater floor and creates a series of rilles arcing through the crater's interior.

Letronne and Gassendi

The horseshoe bay of Letronne crater at the upper left and oatmeal cookie-like Gassendi crater at the lower right are like the cities Paris and London in the Charles Dickens classic, *A Tale of Two Cities*. Like Paris and London, both craters are similar in size and age, but each befell a different fate.

Letronne spans 119 kilometers and is about 3.5 billion years old. Gassendi spans 111 kilometers and is about 3.9 billion years old. Both formed on the edge of a giant impact basin, Letronne on the southern edge of the Procellarum Basin and Gassendi on the northern edge of the Humorum Basin. Their rim-to-rim separation is only 110 kilometers, but that short distance greatly altered their appearance.

Letronne's placement on the tilted edge of the Procellarum Basin allowed the lava flows that formed Oceanus Procellarum to spill over the crater's northern rim and flood the crater's interior. Letronne's higher southern rim cradled the lava flow resulting in a horseshoe bay on the southern Procellarum shoreline. Typically, a crater of Letronne's size possesses a central peak that rises two kilometers above the crater floor. The depth of the basalt covering the interior of Letronne is shown by its twin central peaks protruding only 300 meters above the basalt.

Gassendi formed on higher territory on the northern rim of the Humorum Basin and thus escaped being flooded by Humorum lava flows. But Gassendi was still heavily modified by volcanism. Lava pushed upward through fractures in the crater floor and internally flooded Gassendi. Volcanic uplift further raised the floor and created the intricate fractures of Rimae Gassendi that crisscross the crater floor.

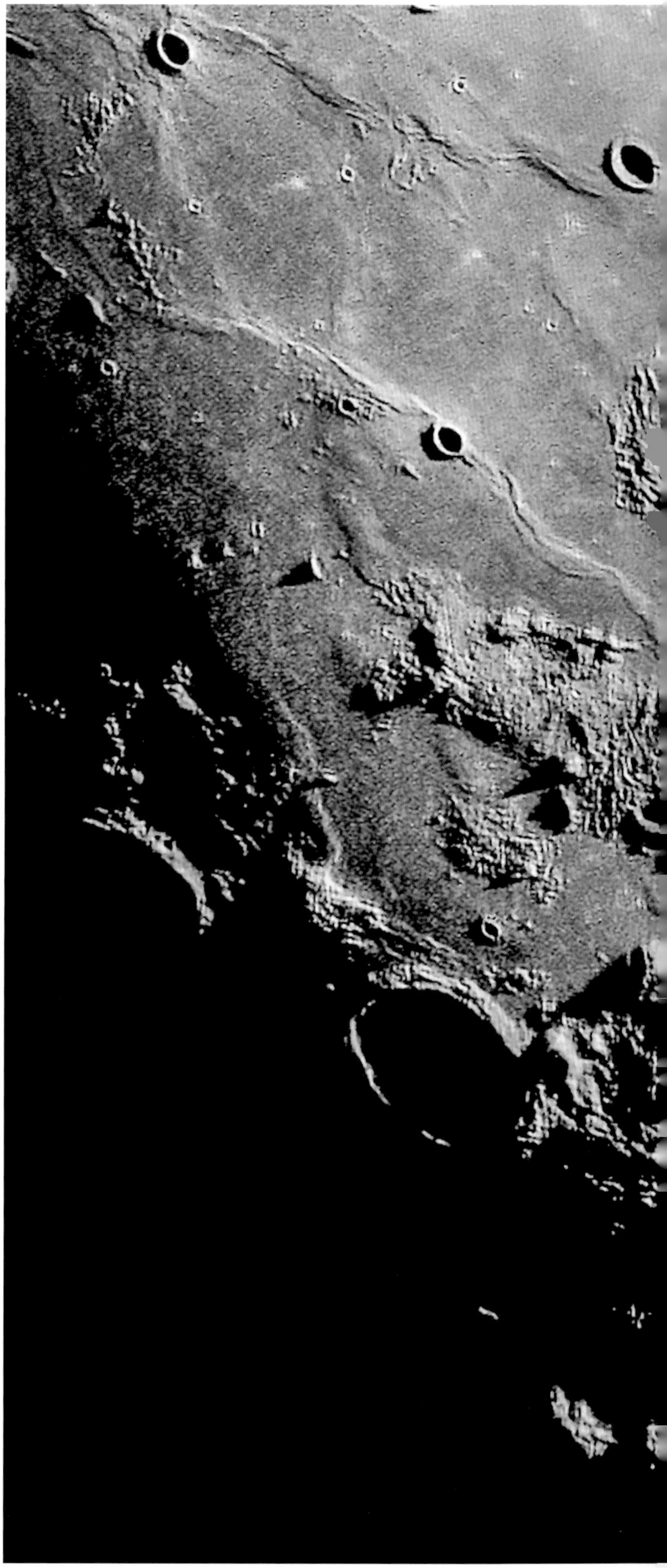

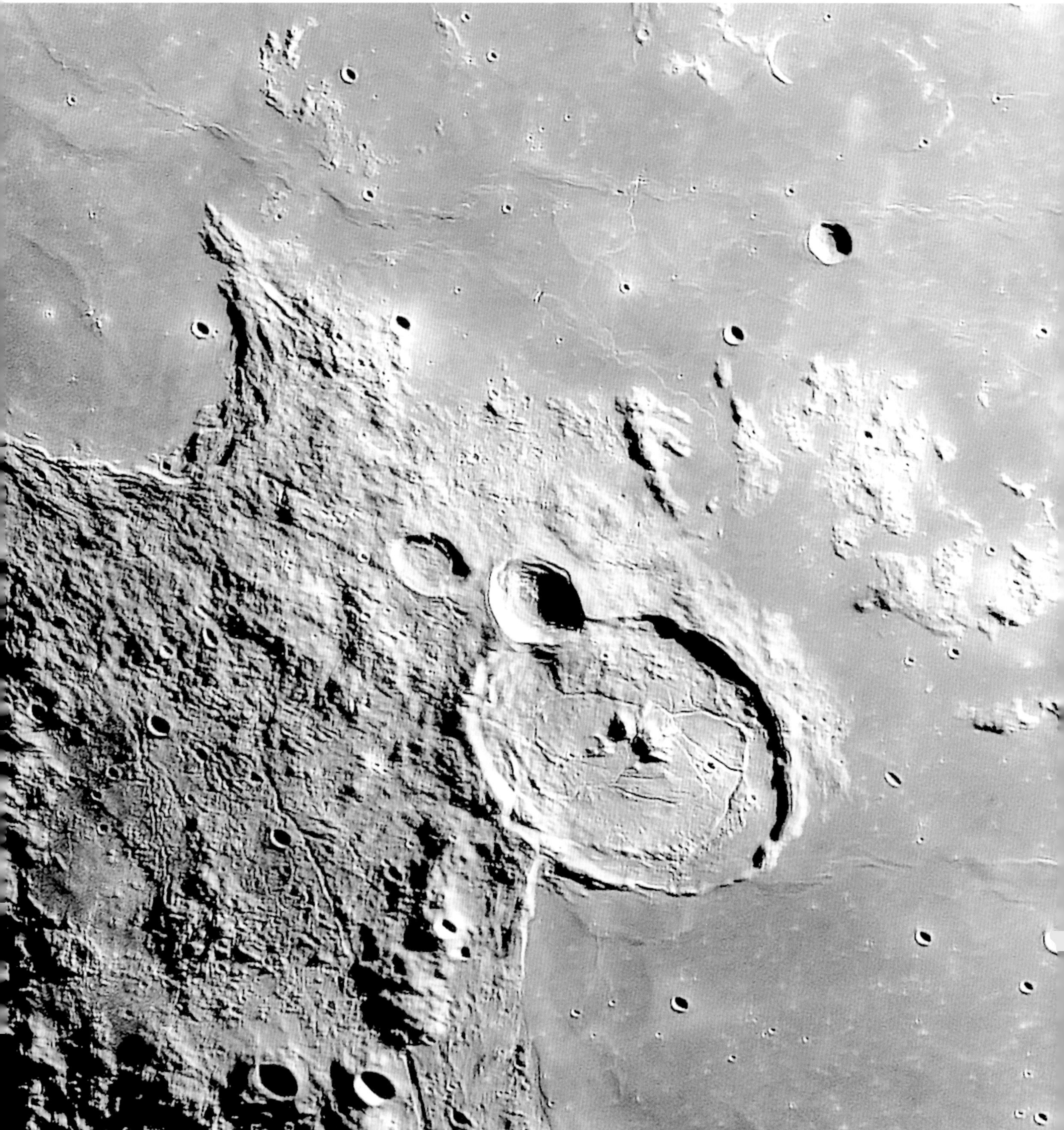

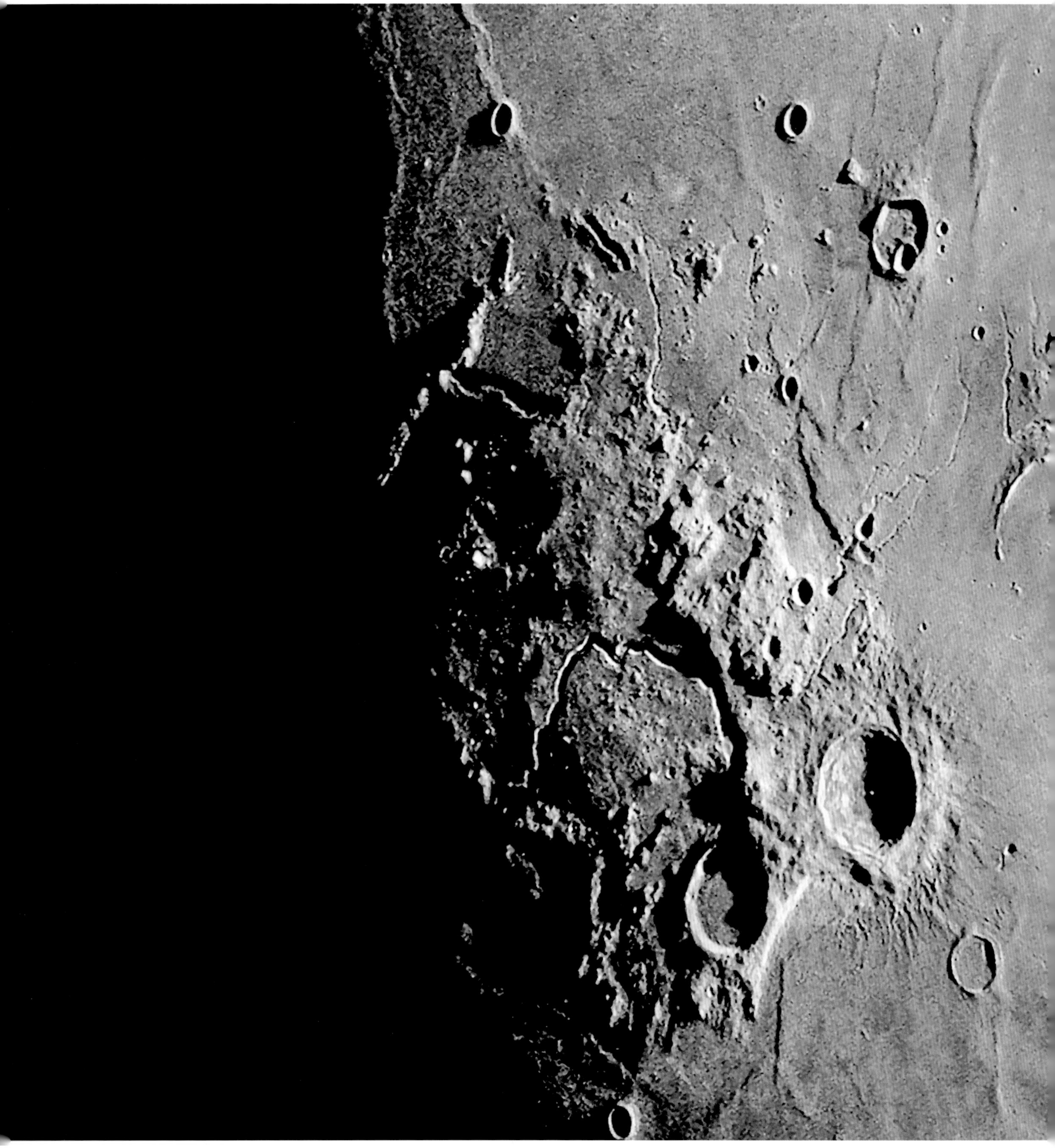

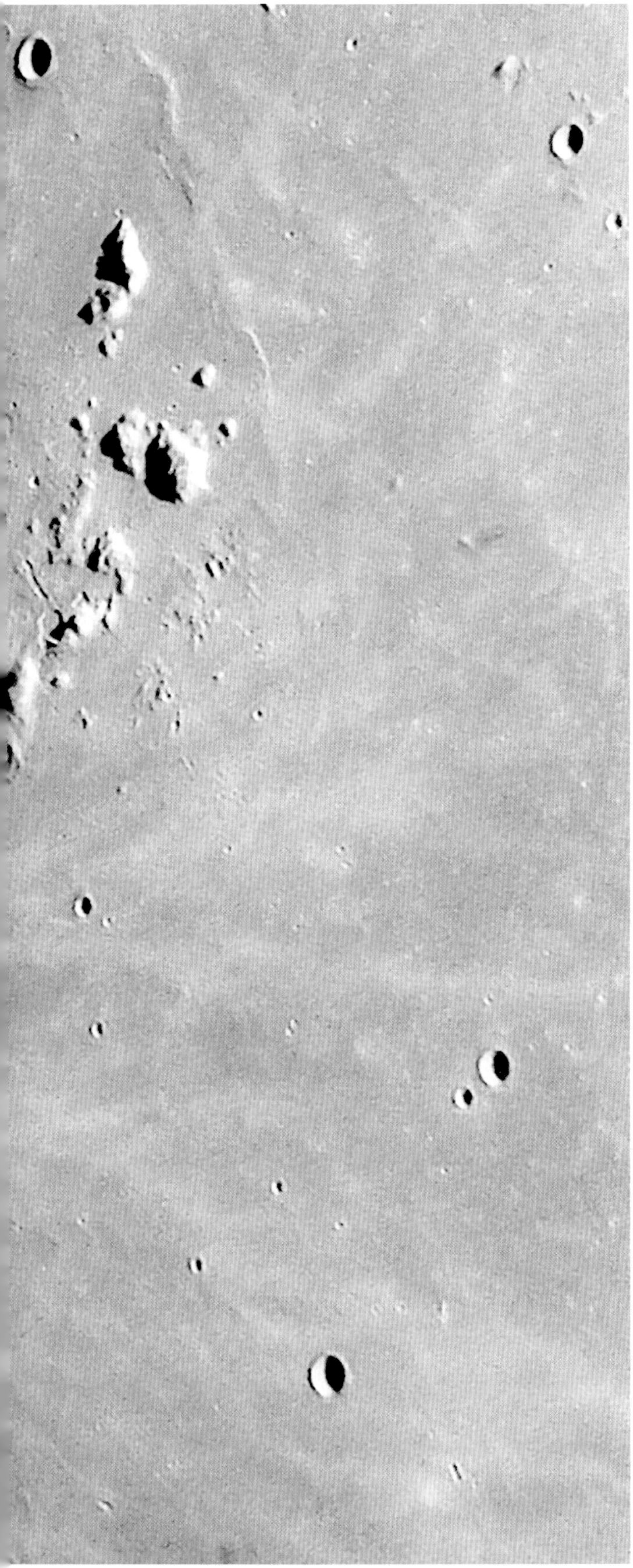

Aristarchus

The volcanic Aristarchus Plateau, spanning 250 by 200 kilometers on northern Oceanus Procellarum, is named after 40-kilometer-wide Aristarchus crater located at the plateau's southeast corner. Aristarchus crater is estimated to be 175 million years old, making it one of the youngest large craters on the near side of the Moon. The plateau is one of the few regions on the Moon that display color. Its greenish-yellow cast is the result of dusting by glassy volcanic ash that erupted from the volcanic pit unofficially known as Cobra Head at the origin of the wide curved volcanic rille commonly called Schröter's Valley. Although it is a volcanic rille that once flowed millions of tons of molten lava onto Oceanus Procellarum, Schröter's Valley is officially designated as a valley, and not a rille, to honor the historic legacy of its name.

Three billion years ago, the one-kilometer-high vertical relief of the Aristarchus Plateau was a volcanic hell that made Pompeii look like a picnic. The squarish Aristarchus region was uplifted by subsurface volcanic pressure. In addition to lava flowing through the snake-like rille of Schröter's Valley, ash spewed from volcanic vents while lava flowed down the multiple branches of Rima Aristarchus and onto the surrounding plains.

The area near the ghost crater Prinz to the right of the plateau was also once volcanically active with the pitchfork-shaped Rima Prinz rille complex spilling lava onto the plains of Oceanus Procellarum. The Montes Agricola mountain chain appears like a wall on the north end of the Aristarchus Plateau. To the east of the plateau, the peaks of the Harbinger Mountains dot the plains of Oceanus Procellarum. The peaks are so named because when they are illuminated by the rising sun, they are a harbinger of sunrise at the Aristarchus Plateau.

Photo Credits

Ballistic Missile Defense Organization 173
China National Space Administration 191, 194, 201
Chinese Academy of Sciences 179, 190
Cornell University 84
Digital Museum of Planetary Mapping 18 (bottom)
Eastman Kodak 119
European Space Agency 174, 176
Firefly Aerospace 203
Indian Space Research Organization 180, 196, 197 (top), 197 (bottom)
Institute of Space and Aeronautical Sciences 169
Intuitive Machines 200, 205
Japan Aerospace Exploration Agency 177, 178, 199
Jet Propulsion Laboratory 57, 64, 65, 86, 88, 117, 129, 132
Kennedy Space Center 136
Memorial Museum of Cosmonautic 74 (public domain)
National Aeronautics and Space Administration 41, 44, 58, 90, 92, 96, 97, 99, 101, 103, 108, 114, 116, 118, 120, 123, 126, 127, 131, 134, 141, 146, 149, 154, 155, 159, 160, 161, 170, 174, 182, 183, 186, 187, 188, 189, front cover, back cover
National Reconnaissance Office 76
Naval Research Laboratory 172
Novosti 61 (bottom)
NPO Lavochkin 112, 139, 143, 157, 198
Pravda 71
Public Domain 15, 16, 17, 18, 22, 24, 26, 29, 32
Rice University 85
RKK Energia 43, 52
RSC Energia 107
Robert Reeves 37, 102, 210–247
Smithsonian Institution 172
Sovfoto 45, 61 (top), 67
SpaceIL 193
Space Technology Laboratories 48, 53, 72
United States Air Force 42, 55, 56, 73, 80, 82, 171
United States Geological Survey 30
University of Arizona 35 (bottom), 39, 40
University of California 35 (top)
USSR Academy of Sciences 79, 105, 110–111, 122, 124, 152, 153, 164–165
Viacheslav Argenberg 94 (CC BY 4.0)

Index

Page numbers in italics refer to photographs.

M

N

O

P